An Introduction to
Biological Control

An Introduction to
Biological Control

Robert van den Bosch
P. S. Messenger
A. P. Gutierrez

Division of Biological Control
University of California, Berkeley
Albany, California

PLENUM PRESS • NEW YORK AND LONDON

Library of Congress Cataloging in Publication Data

Van den Bosch, Robert.
  An introduction to biological control.

  Rev. ed. of: Biological control. 1973.
  Includes bibliographical references in index.
  1. Pest control—Biological control. 2. Insect control—Biological control. I.
Messenger, P. S. II. Gutierrez, A. P. III. Title.
SB975.V36   1982   632'.96   81-21125
ISBN 0-306-40706-X   AACR2

This volume is a revised version of BIOLOGICAL CONTROL, by
Robert van den Bosch and P.S. Messenger, published in 1973
by Intext Press, New York. The original text has been
thoroughly updated for this edition.

© 1982 Plenum Press, New York
A Division of Plenum Publishing Corporation
233 Spring Street, New York, N.Y. 10013

Printed in the United States of America

TO ROBERT VAN DEN BOSCH (1922–1978)

For the legacy of good works in biological control, for aspiring to the highest of ideals in environmental protection, for tenacity in the face of adversity, for faithfulness to human principles beyond the reach of most men, and for the genuine love his bright spirit cast upon those fortunate enough to have known him.

TO P. S. MESSENGER (1920–1976)

For the steady leadership of the Division of Biological Control he provided during difficult times, for his many fine scientific works, for his tenacious struggle to enhance and expand the field of biological control, and lastly for the model he provided for students to emulate.

A. P. Gutierrez

Preface

This volume is a revision of *Biological Control* by R. van den Bosch and P. S. Messenger, originally published by Intext Publishers. In the revision, I have attempted to keep the original theme, and to update it with current research findings and new chapters or sections on insect pathology, microbial control of weeds and plant pathogens, population dynamics, integrated pest management, and economics. The book was written as an undergraduate text, and not as a complete review of the subject area. Various more comprehensive volumes have been written to serve as handbooks for the experts. This book is designed to provide a concise overview of the complex and valuable field of biological control and to show the relationships to the developing concepts of integrated pest management. Population regulation of pests by natural enemies is the major theme of the book, but other biological methods of pest control are also discussed. The chapter on population dynamics assumes a precalculus-level knowledge of mathematics. Author names of species are listed only once in the text, but all are listed in the Appendix. Any errors or omissions in this volume are my sole responsibility.

A. P. Gutierrez
Professor of Entomology
Division of Biological Control
University of California, Berkeley

Acknowledgments

Very special thanks must be given to my colleagues, Professors C. B. Huffaker and L. E. Caltagirone, for the very thorough review they provided and for the many positive suggestions they gave. Dr. Lowell Etzel helped develop the sections on insect pathology, and for this I am most grateful. The tireless efforts of Mr. William Voigt require special mention, as he served in very many capacities, among them illustrator, reviewer, typist, and in general an invaluable assistant. Mrs. Joanne Fox and Ms. Barbara Gemmill remained cheerful while typing the many revisions. Messrs. J. K. Clark (University of California, Davis) and P. F. Daley and F. Skinner (University of California, Berkeley) provided many of the photographs of natural enemies reproduced here. Dr. George Poinar (University of California, Berkeley) provided photography and illustrations for the sections on insect pathology, while Dr. D. F. Waterhouse (Chief, Division of Entomology, CSIRO, Australia) provided the illustrations dealing with biological control of the European rabbit and dung in Australia.

A. P. G.

Contents

Chapter 4

Chapter 5

Chapter 6

Chapter 7

Chapter 12

Chapter 13

1

The Nature and Scope of Biological Control

Biological control is a natural phenomenon—the regulation of plant and animal numbers by natural enemies (biotic mortality agents). It is a major element of that force, natural control, that keeps all living creatures (except possibly man) in a state of balance with their environment. We will use the term *biological control* in our discussion to encompass both the introduction and manipulation of natural enemies by man to control pests* (*applied biological control*) and control that occurs without man's intervention (*natural biological control*).

This volume deals largely with the biological control of insects and weedy plants by insects and pathogens. In a sense, then, the title is inexact, since the full scope of biological control is not covered. But this should not be a major concern because the basic principles of the phenomenon are the same for all groups, and in the applied area of biological control the overwhelming emphasis has been on pest insects and weeds, with insects the principal biological control agents involved (see Figure 1.1).

Biological control relates to those biotic agents that prevent the normal tendency of populations of organisms to grow in exponential fashion and to the mechanisms by which such growth is prevented. This is part of the balance of nature that Darwin described. To gain some insight into the awesome significance of the uninhibited multiplication of an organism, we have only to look to *Homo sapiens* to see the effects of our population explosion on the environment including water and air pollution, decimation of plant and animal life, and destruction of soil fertility. Since insects comprise an estimated 80% (perhaps 1–1.5 million species) of all terrestrial animals,

---

*The term *pest* is used here as a matter of convenience, implying that the species' activities cause "damage" to man. In an ecological sense, the species may be merely filling its evolved niche.

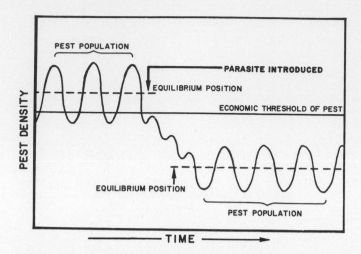

Figure 1.1. Classical biological control resulting in total elimination of an insect pest as an economic problem after the introduction of a parasite. Note that it is not the economic threshold (an artifact of man) that is affected by the introduced parasite, but rather the pests' equilibrium position (long-term mean density) (after Smith and van den Bosch, 1967).

even the partial inhibition of naturally occurring biological control would engender unimaginable consequences. Man might not survive the intense competition for food and fiber he would face or the reduced health he would suffer from the unleashed hordes of insects.

Biological control, then, is of great importance to us and most probably critical to our survival.

Definition of Biological Control

DeBach (1964) discussed the semantics of the term *biological control*. He concluded that the term can refer to a natural phenomenon, a field of study, or an applied pest control technique involving manipulation of natural enemies. In this light, the following definition seems most appropriate since it is simple and yet embraces DeBach's three semantic variations. Biological control is thus defined as "the action of parasites, predators, and pathogens in maintaining another organism's density at a lower average than would occur in their absence" (DeBach, 1964). Close analysis of these few words reveals that they describe a natural phenomenon, denote a field of study, and accommodate the possibility of deliberate natural enemy manipulation. We would find it very difficult to concoct a better definition of biological control.

Some people hold a wider view of biological control that embraces such factors as host resistance, autosterilization, and genetic manipulation of species. These are treated in Chapter 11. But we prefer the narrower concept, first because it is the traditional one, and second because it is neatly delimited by the phenomena of predation, parasitism, and pathogenesis.* We should also note that herbivores are but plant predators, as predation and herbivory are ecologically analogous processes that occur at different trophic levels.

A Comparison of the Biological Control of Pest Insects and Weeds

In principle there is little difference between the biological control of insect pests and weeds. Both involve natural enemies that act to suppress or maintain pest or potential pest species below economically injurious levels. And with each, where natural enemy importation has been employed, the successes have been overwhelmingly against alien pest species. But there are some differences in the biological control of the two pest groups. For one thing, with the plant-feeding insect, a high degree of host specificity, preferably *monophagy*, is an absolute necessity, for there cannot be the remotest chance that the species will develop an affinity for any plant of economic value. Therefore, insects under consideration for importation against weedy plants are subjected to intensive feeding and host preference tests before being cleared for release. There is simply no margin for error in this process, for once a weed-feeding insect is released into the new environment it cannot be called back. On the other hand, with *entomophagous* insects, *oligophagy*, or even *polyphagy* may sometimes be advantageous, and certainly there is no hard and fast stipulation that an imported parasite or predator be narrowly specific. With entomophagous insects the basic concerns are merely that no beneficial species (e.g., the honeybee, lady beetles) be endangered or that hyperparasites (parasites of parasites) be imported.

The need for specificity in weed-feeding insects places a heavy burden of responsibility on everyone involved in the importation process, particularly on those who do the actual testing. In practice, the candidate insect species is first intensively tested in the overseas collecting area for biological characteristics, host plant affinities, and oviposition habits. The plants used in the feeding tests range from wild species related to the host weed through a variety of plants of economic value.

The overseas testing establishes whether an insect species will be passed

---

*Entomologists customarily use the term *parasite* for insects that are parasitic in or on other insects, and the term *pathogen* for microorganisms that cause disease in insects. On the other hand, parasitologists concerned with the medical and veterinary sciences commonly use the term *parasite* to refer to any organism that lives on or in a host, including both microbes and multicellular organisms. To avoid confusion, entomologists often distinguish parasitic insects as *parasitoids*. The distinction is elaborated further in the text. Where we use the terms *parasite*, *parasitism*, and *parasitic*, these are to be understood as referring to insects living in or on host insects. Parasitic microbes that attack and cause disease in insects are called *pathogens*.

on to the domestic quarantine facility for additional intensive study and testing. Then, after this second screening is completed, the data are reviewed by a special committee of experts who make the final decision as to whether field releases shall be made.

The suppression of weedy plants by imported natural enemies differs somewhat from the suppression of insects pests by natural enemies. With insect pests, suppression usually results directly from premature mortality produced by the natural enemy. But with weeds, the role of the natural enemy is more complex; thus (1) it may directly kill its host, (2) it may so weaken or stress the weed that aggressive competitors displace it or make it susceptible to other, preexisting mortality agents, (3) it may impair the reproductive capacity of the weed by destroying its seeds or flowering parts, or (4) its feeding lesions may create an avenue for fatal infection by pathogens. Figure 1.2 illustrates the kinds of physical damage caused by natural enemies of

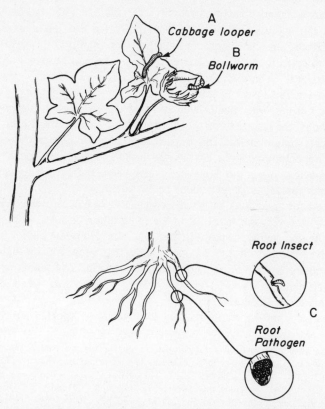

Figure 1.2. Several kinds of plant injury to domestic cotton (A). Leaf feeding. (B) Fruit destruction. (C) Damage to the vascular system from insect feeding or plant pathogen damage.

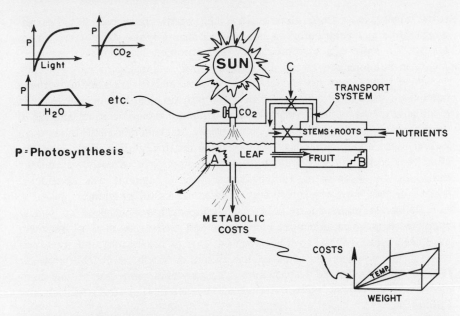

Figure 1.3. A diagrammatic representation of the physiology of a plant attacked by various kinds of natural enemies. (A) Leaf feeding damage, showing reductions in leaf area and wound healing losses. (B) Fruit feeding damage, indicating losses in progeny production (yield). (C) Transport system damage from direct feeding or blockage from pathogens.

plants, while Figure 1.3 illustrates some of the physiological effects of the damage on the plant. The use of domestic cotton in Figure 1.2 merely illustrates the fact that herbivores are beneficial when they help man, and pests when they hinder him.

Despite the technical and subtle mechanical differences just described, biological control of insects and weeds operates under the same broad principles, and the two will be treated together in the following pages.

The Role of Pathogens in Biological Control

Pathogenic viruses, bacteria, fungi, protozoa, and nematodes play important roles in the regulation of plant and insect numbers (Steinhaus, 1949, 1963). A wide range of pest and potential pest species are afflicted by diseases that either prevent them from attaining damaging levels or greatly reduce their potential to cause injury. Among the major pest insect groups, species of Orthoptera, Homoptera, Hemiptera, Lepidoptera, Coleoptera, Diptera, Hymenoptera, and even species in the Acarina (e.g., Tetranychidae) suffer significantly from disease. In other words, microbial parasites are frequently major contributors to naturally occurring biological control. Chapter 5 has a more extensive discussion of this topic.

Man has long recognized the important role of pathogens as controlling agents in nature, and his thoughts long ago turned to ways to deliberately manipulate them. For example, d'Herelle, early in this century, deliberately distributed the bacterium *Coccobacillus acridiorum* d'Herelle in an attempt to initiate epizootics in grasshopper populations. At about the same time, there was also an attempt to induce epizootics of the fungus *Beauveria bassiana* (Basalmo) Vuillemin in populations of the chinch bug, *Blissus leucopterus* (Say), in the midwestern United States. Neither attempt was considered successful, but they were precursors of later successful programs.

Deliberate introductions of exotic pathogens have not been so widely attempted as have introductions of parasitic and predaceous insects. But there have been some successes with pathogens, such as the introductions (inadvertent) of the polyhedrosis virus of the European sawfly, *Diprion hercyniae* (Hartig), from mainland Canada into Newfoundland, and that of the granulosis virus of codling moth, *Laspeyresia pomonella* (L.), from Mexico into California.

But the greatest strides in the manipulation of pathogens have been in their development as microbial insecticides. The first pathogen to be developed, marketed, and utilized in this way was *Bacillus popilliae* Dutky, the famed milky disease of the Japanese beetle, *Popillia japonica* Newman. An even more outstanding microbial insecticide has been *Bacillus thuringiensis* Berliner, which in its various strains is an effective killer of a broad spectrum of lepidopterous pests. *B. thuringiensis* is under commercial production both in the United States and aboard and must be considered a major modern insecticide. It is particularly desirable material because of its high degree of selectivity among arthropods and its complete safety to the environment and warm-blooded animals.

Several viruses have also been effectively used as microbial insecticides. The viruses are even more selective than *B. thuringiensis* and thus would seem to be of high promise as safe insecticides. However, since they are viruses, they are being exhaustively tested for possible pathogenicity to other groups of animals. This has slowed their federal registration for insecticidal use, increased their developmental costs, and delayed their commercial exploitation. However, it seems only a matter of time before the registration protocol is developed for viruses, permitting such promising ones as the polyhedrosis viruses of bollworm, *Heliothis zea* Boddie, cabbage looper, *Trichoplusia ni* Hubner; beet armyworm, *Spodoptera exigua* Hubner; and the granulosis virus of codling moth, *L. pomonella*, to be brought into widespread use. A more detailed treatment of this topic is given in Chapter 5.

Summary

This chapter has introduced the major areas of biological control covered in this text, but a larger treatment is reserved for other chapters. Natural control achieved by natural enemies is a process that regulates populations of most species—except man—and biological control is its scientific application.

References

DeBach, P. (ed.) 1964. *Biological Control of Insect Pests and Weeds*. Chapman & Hall: London. 844 pp.

Smith, R. F., and R. van den Bosch. 1967. Integrated Control. In: W. W. Kilgore and R. L. Doutt (eds.) *Pest Control—Biological, Physical, and Selected Chemical Methods*. Academic Press: New York. pp. 295–340.

Steinhaus, E. A. 1949. *Principles of Insect Pathology*. McGraw Hill: New York. 757 pp.

Steinhaus, E. A. (ed.) 1963. *Insect Pathology—An Advanced Treatise*. Academic Press: New York. Vol. 1, 661 pp. Vol. 2, 689 pp.

2

The Ecological Basis for Biological Control

Biological control is a natural phenomenon that, when applied successfully to a pest problem, can provide a relatively permanent, harmonious, and economical solution. But because biological control is a manifestation of the natural association of different kinds of living organisms, i.e., parasites and pathogens with their hosts and predators with their prey, the phenomenon is a dynamic one, subject to disturbances by other factors, to changes in the environment, and to the adaptations, properties, and limitations of the organisms involved in each case (Huffaker and Messenger, 1964).

In order to understand the potential and the limitations of biological control, and especially in order to carry out competently a program of biological control, an awareness of the ecological basis of the phenomenon is essential. Three interrelated concepts that must be understood are (1) the idea of discrete populations and communities, (2) the balance of nature, and (3) the natural control of numbers.

Biological control is a manifestation of the association of different, interdependent species in nature. But species exist as groups of like individuals. These interbreed, reproduce, and die. By reproducing they maintain themselves as a group, which at a local level is called a population (Boughey, 1971).

A population changes in size, that is, in the number of individuals it contains, according to whether environmental (biotic and abiotic) circumstances favor the production of more or less individuals than the number dying in a given interval of time. Migration of individuals into or out of the local population also affects population size.

Another important feature of a population is its age structure. *Population age structure*, a term often simplified to *population structure*, means the age pattern distribution of the individuals in the population. The population struc-

9

ture may assume many patterns that depend on the past history of factors affecting the population. At one extreme, all members of the population at any one time may be approximately the same age or in the same stage of development. At the other extreme, individuals of all ages occur together. In the first case, the life cycles of all population members are synchronized with each other, a situation often brought about by annual climatic cycles or by some facet of the species' biology, such as quiescent periods. In the second case, generations are not synchronized but strongly overlap, a pattern commonly found in populations of short-lived insects with many generations per year or in populations of insects that display continuous reproductive activity uninterrupted by seasonal climatic cycles (e.g., some aphids in warm climates). Through time, the populations are composed of differing proportions of individuals of widely varying ages, a situation often induced by age-specific mortality factors (e.g., parasitism). This is in sharp contrast to the *stable age distributions* in some human populations, where the proportions in the different age groups have tended to remain reasonably constant.

Age structure is important in respect to host populations in which only one or two stages of development are utilizable by a particular natural enemy. Close synchronization between natural-enemy and host life cycles must occur if successful control of the host is to be achieved. Such is the reason why the encyrtid parasite *Metaphycus helvolus* (Compere) is an effective control agent of black scale, *Saissetia oleae* (Olivier), on citrus in coastal southern California, while it is much less effective against the same pest in interior southern and central California. In the former area, the scale population structure includes all ages at once because of the lack of a synchronizing feature in the life cycle; in the latter areas, the scale population structure is restricted to only one or two of the several age classes at any given time, so that for certain periods of the year no suitable host stages are available for the parasite to attack. The parasite population is then unable to reproduce and as a result diminishes in numbers.

Age structure is also important in the population dynamics of an insect, since it often reflects the growth phase of the population. When a population is young and is beginning to increase in numbers, its age structure includes relatively many of the younger ages, less of the intermediate ages, and few adults. When a population is mature and no longer increasing in number, perhaps because of intraspecific crowding, it is composed of relatively less young and more adults. Chapter 7 presents a review of methods used to analyze and estimate the impact of various biotic and abiotic factors causing population change.

Populations are also dynamic with regard to geographic distribution. They tend to spread in space until some limiting environmental condition is encountered, such as a geographic barrier (e.g., a coast, mountain range, or desert boundary), or an environmental limitation, such as the absence of a required resource (e.g., a necessary food organism or a specific soil habitat).

Populations do not exist in isolation; they occur in habitats in association with other species. Such assemblages of species populations constitute *communities*. To the degree that certain species are consistently associated with each other and can be recognized—particularly the case with certain characteristic plants—we can distinguish discrete communities (Odum, 1971). For example, we can describe a pine forest community and can usually find certain species of insects and other animals associated with the dominant pine species.

In communities, we can distinguish trophic or nutritional associations between interacting species. Thus, we recognize *primary producers* (i.e., green plants), *primary consumers* (i.e., herbivores), *secondary consumers* (i.e., carnivores), *decomposers*, and *scavengers* (Odum, 1971). *Food chains* can commonly be discerned, wherein a given plant species is consistently fed on by a defoliating insect, for example, which in turn is fed upon by certain bird species. This is a three-step chain, but if the insectivorous bird population is preyed upon by a hawk, we recognize a fourth trophic step. Such food chains, because of simple material and energy losses along the way, are not endless but typically occur in links of four to six. Figure 2.1 shows the flow

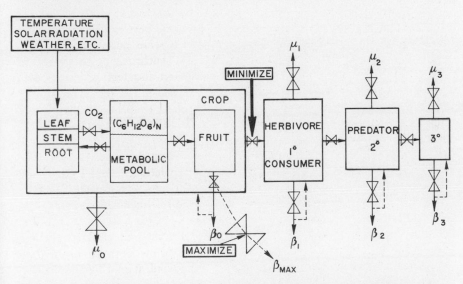

Figure 2.1. The production (photosynthesis) and flow of biomass and energy up the food chain to deaths (μ_i) and births (β_i). The values imply energy flow, while the heavy arrow implies pesticide use or biological control introductions sufficient to limit the flow of energy up the food chain.

and dissipation of energy up the food chain in a natural ecosystem. In agriculture, pesticides and preferably biological controls (heavy arrow in the figure) are used to shunt the energy into a harvestable crop.

Where food chains branch or join together, as generally happens in complex communities, the complex of trophic paths is referred to as a *food web*. Such webs can be discerned where one herbivore feeds on more than one plant species, or where several bird species include one defoliating insect species in their diet. Figure 2.2 illustrates such a food web.

Thus we encounter on a universal scale such trophic interactions as *phytophagy*, the consumption of plants or plant parts by herbivorous animals, and *carnivory*, the consumption of or nutritional dependency on insects, particularly phytophagous ones, of certain animal species. Thus we work with such entomophagous animals as insect-eating mammals (shrews, mice), insectivorous birds, fish, amphibians, and, particularly, entomophagous insects.

It is customary in biological control work to describe the species of animals and plants that live at the expense of other plants and animals as *natural enemies* of the latter. Any given species in a community, with few exceptions, is attacked and fed upon by one or more such natural enemies, an indication of the tremendous potential of biological control.

All organisms are capable of increasing in numbers through processes of reproduction. Most insects are particularly notable for their high potential

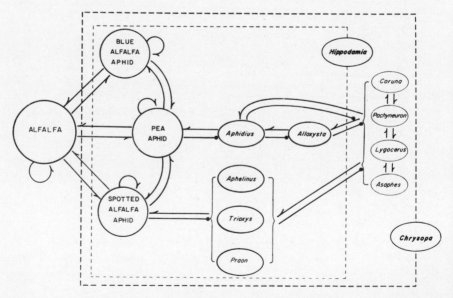

Figure 2.2. A diagram of species interactions on several trophic levels in alfalfa. An arrow implies an effect of one trophic level on another one, no arrow implies no effect, and the "self" arrow implies density-dependent self-limitation.

rates of numerical increase because of their relatively very high fecundities and short life cycles. These species are commonly called *r* strategists, and some good examples are found among parthenogenetic aphids. Others reproduce much more slowly (*k* strategists), are relatively long-lived, and are food-limited (e.g., the tsetse fly and the elephant).* In all cases, the reproductive potential of all species is greater than the replacement rate. But the fact is that most organisms, including the rapidly breeding insects, do not increase over successive generations or for prolonged periods. On the contrary, they increase only periodically and to limited extent as a consequence of natural controls present in their environments (Solomon, 1949). The reproductive strategy employed by the species has evolved to accommodate the various facets of the physical and biotic environment (i.e., ecological *fitness*).† This kind of relationship was shown for a plant–aphid–parasite relationship by Gilbert and Gutierrez (1973) and Gutierrez *et al.* (1979) who showed that each species co-adjusted its reproduction rate so that each trophic level took only the optimal amount from the lower level sufficient for each species to maximize its *fitness*.

Natural controls generally limit population numbers. Such checks to numerical growth include limited resources (food, space, shelter), periodically occurring inclement weather or other hazards (heat, cold, wind, drought, rain), competition among themselves or from other kinds of animals, and natural enemies (predators, parasites, pathogens) (Figure 2.3). This last category is particularly important for many insect species, for while resources may rarely appear to be in short supply, weather may be favorable for long periods, and competitors scarce or absent, natural enemies are almost universally present, often significantly so. Commonly, economic problems arise when a pest species is accidentally introduced to a new area without its adapted natural enemies. But even pest species having their natural enemies present may still cause economic damage because our economic tolerances of their numbers may be much lower than their evolved average steady densities. An excellent example of such a relationship is that between sylvan cotton (*Gossypium hirsutum* L.) and its evolved herbivore, the cotton boll weevil (*Anthonomus grandis* Boh.) (Gutierrez *et al.*, 1979). This conflict has led to the treadmill of massive insecticide use in domesticated cotton.

Probably every insect population in nature is attacked to some degree by one or more natural enemies. Indeed, referring just to entomophagous insects alone, Clausen (1940) stated that probably every phytophagous insect species is attacked by one or more parasitic or predatory insect species. But other predatory animals also act as stabilizing natural control agents of insect

---

*The classification of *r* and *k* strategists is naive, but is used here for historical reasons.
†Fitness as defined by R. A. Fisher is survival rate × reproduction over some effective time period, i.e., a season or generation (see Gilbert *et al.*, 1976).

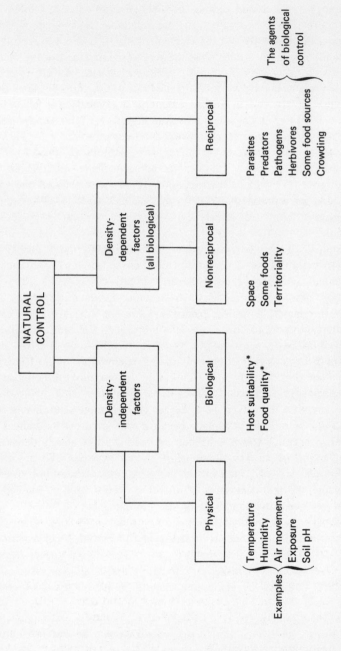

Figure 2.3. The major components of the natural control of population numbers. (*)Suitability and quality of food can be density-dependent regulation factors if some density-related factor is affecting them.

populations, for example, some birds, certain mammals, toads, frogs, and lizards (Fujii *et al.*, 1978).

The result of natural control is the regulation of numbers, preventing the population from becoming too high or relaxing certain suppressive influences when the population becomes low. The occurrence of this, the long-term maintenance of a population at a characteristic level of abundance relative to other organisms in the community, is a demonstration of the balance of nature (Huffaker and Messenger, 1964). The characteristic level of abundance of a species is not static, but rather implies that in the long-run, a population fluctuates about some average that enables it and other species with which it interacts to survive. The mechanisms and interactions between the population and its environment that bring about the relative balance of numbers constitute the natural control of populations. Natural control includes the collective forces of the environment that serve to hold a given population in check against its own capability for numerical growth. As such, natural control (especially through mortality factors or influences and in some cases through factors acting on natality or reproductive capability) includes (1) climatic factors, such as excessive heat, cold, or aridity, (2) disappearance or deterioration of food resources, and (3) the action of competing species and natural enemies. This whole field of study is often referred to as *population dynamics* or *population ecology* in this narrow sense.

It is useful to distinguish between environmental factors such as weather, which act as mortality agents at intensities unaffected by the size of the population, and those whose intensities of action vary with the abundance of the species in question, as seen with competition for food or when predators consume proportionately more when the prey are abundant than when they are scarce. The former type of limiting factor is called a *density-independent* mortality factor, and the latter, a *density-dependent* mortality factor (Smith 1935). The relationship between prey densities and the number of prey attacked is called the *functional response*. (For a more detailed discussion see Chapter 7.) Natural enemies have the capability of acting as density-dependent mortality factors, though they may not act this way, depending on their own environmental and behavioral limitations. All natural enemies used in or contributing to successful biological control programs act in this way.

Density-dependent factors may also be classified according to whether they vary in numbers (or magnitude) in response to changes in host numbers (*reciprocal action*) or whether their numbers (or magnitude) remain fixed even though some resource may change (*nonreciprocal action*). Parasitoids and predators are examples of the former, since they commonly increase in numbers when their hosts or prey become numerous and decrease as the hosts or prey are suppressed. That is to say, the enemies "control" their hosts, and the hosts "control" the enemies. Space is an example of a nonreciprocal factor since it does not wax and wane as the user population rises or falls, but the crowding between individuals in that space as density increases may have a regulating effect on the population's growth. For example, crowding is known to have a very pronounced influence on the production of migratory forms in some aphids at densities well below levels that would greatly affect the host plant population (Johnson, 1969). Space can "control" the numbers of the users, but the users do not always alter the amount of space present (e.g., nesting sites).

For insect populations to remain in relative numerical balance within their normal communities for substantial periods of time, it appears to us necessary that there be one or more intrinsic or extrinsic density-dependent agents that affect mortality, natality or dispersal of such populations. Significantly, mortality agents being responsive to increases in the density of the population in question serve as regulators to check increase; as the population density declines, the regulative action of these agents moderates, allowing the population to rise again. In most species, crowding increases mortality, causes dispersal, and reduces fecundity. Using control-system terminology, the insect population density is regulated by means of a negative-feedback control mechanism, much like a governor on a steam engine or a thermostat in a heating system (Nicholson, 1954).

The existence of density-dependent regulation of insect population densities is now widely accepted and forms the basis for much of the theoretical and applied work in insect population ecology, including biological control. Biological control both in undisturbed nature and in highly managed agricultural environments (*agroecosystems*) is a demonstration of this contention. When an insect population is maintained at a characteristic level of abundance by the effects of natural control agents (including both density-dependent and density-independent mortality agents), the contribution by natural enemies to the total mortality occurring in any generation is usually substantial. Furthermore, the great majority of insects are rare, such that of the one million or more insect species, perhaps only from 10,000 to 30,000 species feed on crops, forest, pasture, or livestock, menace our health, comfort, or possessions, and are abundant enough to be recognized as economically important. Therefore, when insect species invade new geographic regions as an accidental result of man's commercial activities, for example, they increase many times to extraordinarily high numbers mainly because they have escaped the controlling influences of their customary natural enemies. Overall generation mortality is greatly diminished, while the reproductive rate of survivors remains high until limited by some other factor(s). Populations of such invaders therefore begin to increase in numbers at an exponential rate, and we soon have a population outbreak. In such cases, the characteristic abundance of the species is often quite high, and the population may fluctuate enormously. These populations are often as not suppressed by limited food supply or adverse weather.

When such an invading insect is injurious, pest outbreaks occur, and control efforts are required. It is very logical with such pests that efforts be directed to the search for and colonization of any adapted natural enemies that remain behind in the native home of the invading species. Virtually all successful classical biological control programs to date have resulted from the reassociation of invading pests of foreign origin with their adapted natural enemies (DeBach, 1964).

Biological Control of Native Species

Not all insect pests are of foreign origin; some, including a number of the most serious, are native species. The very important cotton bollworm, *Heliothis zea* Boddie, and the lygus bug, *Lygus hesperus* Knight, are em-amples of pests native to the western United States, as are the apple maggot, *Rhagoletis pomonella* Walsh; pecan aphid, *Monellia costalis* Fitch; and spruce budworm, *Choristoneura fumiferana* (Clemens), native to the northeastern and north central parts of the United States and Canada. The question is: Can classical biological control provide any help against such pests as these?

The case for biological control of native pest species is technically more complicated than it is for the invading foreign pest. A native species must be presumed to have already associated with it an assortment of adapted natural enemies, and after investigation this is usually found to be the situation. But the agricultural practices of the growers (e.g., pesticide use) for the crop under attack by the native pest tend in many instances to unduly favor the increase in numbers of the pest. Such practices also often interfere with the efficacy with which native natural enemies exert their share of the natural control of such pests (Figure 2.4). Biological control of such native pests can then assume several routes: (1) the introduction of natural enemies of foreign origin that are associated with related insect species, (2) the modification of agricultural and other practices with the intention of enhancing native natural enemy action, or (3) the employment of other pest control techniques, chemical control among them, to bring about the *integrated control* of such pests.

Some invading pests have been controlled by natural enemies derived from related host species. For example, in Hawaii, the importation of the Queensland fruit fly parasite, *Opius tryoni* Cameron, provided an important biological control agent against the Mediterranean fruit fly, *Ceratitis capitata* (Wiedemann). This has led to hope that native pests might be similarly controlled. However, parasites, which experience shows are most often the best biological control agents, are in the great majority of cases highly host specific. That is, their natural adaptations to the parasitic mode of life are narrowly limited to the particular host species with which they have evolved. Past efforts to use imported parasites from related hosts on native host species have only rarely succeeded. This is not true for predators. However, the

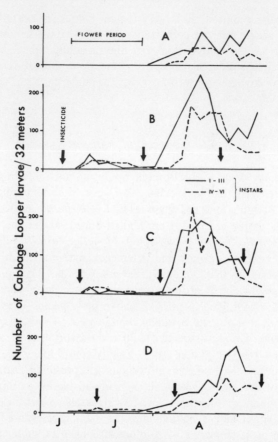

Figure 2.4. Observed phenologies and patterns for cabbage looper larvae in three pesticide application experiments (B, C, and D). Control is (A). Arrows indicate insecticide application dates (adapted from Gutierrez *et al.*, 1975).

knowledge that some parasites can attack more than one host species does not indicate that these species should be neglected in the development of future biological control programs.

Alterations of normal agricultural practices in some cases has favored the increase of native natural enemies. Although not commercially used, the strip-cropping of alfalfa protects the numerous beneficial species associated with this plant, resulting in better control of the phytophagous pest species in alfalfa and in nearby cotton fields (e.g., lygus bug and bollworm). The addition of shelter boxes to tobacco fields increases the numbers of predatory wasps that help control tobacco worms. The planting of wild blackberry plants near grape wineyards provides an alternate host for an egg parasite of the grape leafhopper, *Erythroneura elegantula* Osborne. Often these practices are not used because the inconvenience perceived by the farmers is greater than the perceived benefit.

The inclusion of biological control agents in the integrated control programs against some of our most serious native pests has, however, proven to be of much value but also of such complexity that we present a fuller treatment of integrated control later in this book.

References

Boughey, A. S. 1971. *Fundamental Ecology*. Intext Educational Publishers: New York. 222 pp.

Clausen, C. P. 1940. *Entomophagous Insects*. McGraw-Hill: New York. 688 pp.

DeBach, P. 1964. *Biological control of insect pests and weeds*. Chapman & Hall: London. 844 pp.

Fujii, K., P. M. Mace, and C. S. Holling. 1978. A Simple Generalized Model of Attack by Predators and Parasites. Inst. Res. Ecol. Univ. Brit. Columb. Rept. #R-16.

Gilbert, N. E. and A. P. Gutierrez. 1973. A plant–aphid–parasite relationship. *J. Anim. Ecol.* 42:323–340.

Gutierrez, A. P., Y. Wang, and R. Daxl. 1979. The interaction of cotton and the boll weevil (Coleoptera: Curculionidae)—A study of coadaptation. *Can. Entomol.* 111:357–366.

Gutierrez, A. P., L. A. Falcon, W. Loew, P. A. Leipzig, and R. van den Bosch. 1975. An analysis of cotton production in California: A model for Acala cotton and the effects of defoliators on its yield. *Environ. Entomol.* 4:125–136.

Huffaker, C.B., and P. S. Messenger. 1964. The concept and significance of natural control. In: P. DeBach (ed.) *Biological Control of Insect Pests and Weeds*. Chapman & Hall: London. pp. 74–117.

Johnson, C. G. 1969. *Migration and Dispersal of Insects by Flight*. Methuen: London. 763 pp.

Nicholson, A. J. 1954. An outline of the dynamics of animal populations. *Austral. J. Zool.* 2:9–65.

Odum, E. P. 1971. *Fundamentals of Ecology*. 3rd ed. Saunders: Philadelphia. 574 pp.

Smith, H. S. 1935. The role of biotic factors in the determination of population densities. *J. Econ. Entomol.* 28:873–898.

Solomon, M. E. 1949. The natural control of animal populations. *J. Anim. Ecol.* 18:1–35.

3

The History and Development of Biological Control

The purposeful control of insect and weed pests by biotic agents is a comparatively modern development, having become an effective technique in pest control only since about 1890. However, there are antecedent historical events that trace the evolution of some of the fundamental concepts in the development of biological control, and several of these events show the remarkable and perceptive insight of man into the workings of nature. Without these pre-nineteenth-century discoveries and conceptualizations, modern environmental science, to which biological control has made substantial contributions, would very likely have been much delayed. These discoveries and concepts include, among others, those of the balance of nature; population growth and limitation; natural control of numbers; the symbioses among different species, particularly those of plants, animals, and their natural enemies; and the roles such natural enemies play in the determination of abundance.

The history of applied biological control to a large degree reflects our increasing knowledge of ecology. Indeed, particularly during the late nineteenth century, the ideas and concepts underlying biological control contributed in important ways to the developing theories and principles of ecology and reinforced the practical formulation of biological control of pests. This is not surprising, since biological control is in its essence an ecological phenomenon, and in its practice is an example of "applied" ecology.

A number of books on entomology or biological control include useful sections of the historical features of this subject. These include, in chronological order: *A History of Entomology* by E. O. Essig (1931), *The Biological Control of Insects* by H. L. Sweetman (1936), *The Principles of Biological Control* by the same author (1958), *Beneficial Insects* by L. A. Swan (1964), the very comprehensive *Biological Control of Insect Pests and Weeds* edited by P. DeBach (1964), the latter *Biological Control by Natural Enemies* by

DeBach (1974), and *The History of Biological Control* by K. S. Hagen and J. M. Franz (1973). Besides consulting much of the original literature, we have gleaned some of the following material from one or more of these sources, leaning heavily on the detailed chapter titled "The Historical Development of Biological Control," prepared by our colleague R. L. Doutt (1964), contained in the collection edited by DeBach mentioned above. We must acknowledge three further references, particularly for facts concerning details of the natural enemies used or of the projects carried out in the United States since the beginning of the twentieth century. These are: *Entomophagous Insects* (1940), *Biological Control of Insect Pests in the Continental United States* (1956), both by C. P. Clausen, and *Theory and Practice in Biological Control* by C. B. Huffaker and P. S. Messenger (1976).

Ancient Origins

The idea that insects could be intentionally used to suppress populations of other insects is an ancient one. Our earliest knowledge of this practice seems to have originated with the Chinese and, not surprisingly, involved the use of predators—in this case predatory ants—to control certain insect pests of citrus. Indeed, this agricultural practice has descended through the ages, continuing even into modern times in the Orient where citrus growers maintain and sometimes even purchase colonies of the predatory ant, *Oecophylla smaragdina* Fabricius to place in orange trees to reduce the numbers of leaf-feeding insects (McCook, 1882; Clausen, 1956).

This use of predatory ants no doubt derived from the obvious carnivorous habit in this group of insects. Our ancestral farmers must have been aware of the foraging behavior of such forms as the voracious army ant and the carrying off of soft-bodied grubs and caterpillars by the ubiquitous trail-making ant species *Lasius* and *Formica*. But the practical use of ants by the ancient agriculturalists to control pests was not accomplished without some considerable ingenuity. Not all ants are predatory. Indeed, medieval date growers in Arabia seasonally transported cultures of predatory ants from nearby mountains, where presumably they occurred naturally, to the oases to control phytophagous ants that attacked date palm. This practice constitutes the first known example of man's movement of natural enemies for purposes of biological control. It also bears testimony to the ability of the medieval Arabian date growers to distinguish among species of the ant family on the basis of their food habits. And from what we know today about the habits of certain predatory ants (for instance, the Argentine ant, *Iridomyrmex humilis* Mayr) that actually protect certain kinds of phytophagous insects such as aphids, soft scales, and mealybugs from their natural enemies, we must admire still further the insight displayed by these early practitioners of biological pest control.

Awareness of the possibilities for the use of parasitic insects in combating pests was slower in development, most likely because of the much more subtle and cryptic nature of the parasitic habit in insects. Insect parasitism was first recognized in Italy in the seventeenth century by Vallisnieri, who noted the unique association between the parasitic wasp, *Apanteles glomeratus* (L.), and the cabbage butterfly, *Pieris rapae* (L.) (Doutt, 1964). In the early decades of the eighteenth century, there appeared an increasing number of reports referring to the parasitic habit among insects, but the idea that such natural enemies could be used in a practical sense to control pests was not put into practice until the following century.

The first suggestions that biological control through parasites might be a practical solution to pest problems were European in origin. Erasmus Darwin (1800) noted the destruction of cabbage butterfly caterpillar infestations by the "small ichneumon-fly which deposits its own eggs in their backs" (Doutt, 1964). The gathering and storing of parasitized caterpillars in order to harvest parasite adults for later release was proposed by Hartig in 1826 in Germany (Sweetman, 1936). In France, Boisgiraud in 1840 collected and liberated large numbers of predatory carabid beetles, *Calosoma sycophanta* (L.), to destroy leaf-feeding larvae of the now famous gypsy moth, *Porthetria dispar* (L.). In Italy, Villa in 1844 proposed and later demonstrated the use of predatory insects, in his case carabid and staphylinid beetles, to destroy garden pest insects.

In Europe, the initial applications of biological control concerned the use of locally derived native parasites and predators to control local infestations of pests. There was no suggestion or effort to bring such natural enemies from distant places to control local pests. This was very likely because most European agricultural or garden pests were considered native to the localities infested, crops were rarely ravaged by pest outbreaks of foreign origin, and the idea that insect parasites and predators from foreign lands might prove useful in combating pest infestations probably never came to mind—nor is it given the attention it deserves there even today.

North American Beginnings

In America, with the development and rapid spread of agriculture from the eastern seaboard to the expanding western frontier in the early nineteenth century, came the increasing occurrence of insect pest outbreaks resulting mainly from insect species of foreign origin. For example, the highly damaging wheat midge, *Sitodiplosis mosellana* (Gehin), was known to have come to America from Europe, and Asa Fitch (1809—1879), State Entomologist for

New York, conjectured that its persistence at injurious levels was due to the absence of its normally restrictive natural enemies that in Europe, he felt, served to keep it at lower, less harmful numbers. In 1855 Fitch proposed the importation of parasites from England to bring about a reduction in abundance of the midge. While this pioneering proposal came to no immediate practical end, the idea soon received support from other foresighted entomologists, including D. J. S. Bethune in Canada and Benjamin Walsh (1808–1870) in Illinois.

In 1870 parasites of the plum curculio *Conotrachelus nenuphar* (Herbst), a native pest species, were distributed from one part of Missouri to another for control purposes by State Entomologist C. V. Riley (1843–1898). In 1873 Riley arranged the first international shipment of a natural enemy in the transfer of the predatory mite *Tyroglyphus phylloxerae* Riley to France from North America for possible control of the grape phylloxera [*Phylloxera vitifoliae* (Fitch)], a native North American pest accidentally introduced into Europe in the early nineteenth century.

In 1879 Riley was appointed Chief Entomologist for the U. S. Department of Agriculture, Washington, D.C. Soon thereafter (1883), he directed the importation of the internal parasite of the cabbage butterfly from England to America. This introduction was successful, and *A. glomeratus* eventually became well distributed throughout the eastern and mid-western states. However, it did not become a very effective biological control agent and received little recognition.

California Origins

The development of biological control in America remained an eastern and midwestern endeavor until the early 1880s, when the "movement" found support in the burgeoning agricultural enterprises of California. During the period between 1840 and 1870, California agriculture, under the stimulus of a mild climate, exceptionally fertile soils, and abundant water for irrigation, expanded at an extraordinary rate. A multiplicity of new crops resulting from importation of seeds, seedlings, and cuttings of the best varieties of fruit and nut trees, vines, field crops, and ornamental plants from many countries enabled California to become one of the world's major crop producing regions. But unfortunately, along with the importation of all these plants came an increase in insect and disease problems. It is therefore not surprising that as early as 1881 proposals for introducing natural enemies for pest control purposes were submitted by several California horticulturalists. But it was not until the late 1880s that the first planned, successful project in biological control took place involving the serious citrus pest known as the cottony-cushion scale, *Icerya purchasi* (Maskell).

The Cottony-Cushion Scale Control Campaign

The biological control project against the infamous cottony-cushion scale in California not only constitutes the first truly successful example of the use of this pest control technique in the world, but also serves as a classic example, exhibiting all of the basic features characteristic of this method of pest control. The project is worth exploring in some detail.

The cottony-cushion scale, a destructive pest of citrus, pear, acacia, and other plants, was first recorded in California in the town of Menlo Park in 1868, where it was noted infesting acacia plantings in a horticultural nursery. From the original nursery infestation it soon spread to nearby ornamental species, including citrus trees. At this time, the young, burgeoning California citrus industry was concentrated mainly in the Los Angeles area some 400 miles south of Menlo Park. But within three or four years of its original invasion in the San Francisco Bay area, the pest was unwittingly carried to Southern California on infested lemon stock. It became established in the Los Angeles area prior to 1876, when it was documented as well spread through the various citrus groves of what is now the downtown area of that city. Soon thereafter, it was found in San Gabriel Valley citrus groves, some 10 miles to the east, and in the newly developing groves of Santa Barbara, 100 miles to the northwest. By 1880 the pest had become distributed throughout California, damaging citrus trees wherever they were grown.

Specimens of the cottony-cushion scale were sent to C. V. Riley in 1872, while he was still State Entomologist for Missouri. He suggested that the new pest may have come from Australia, since he knew that much citrus nursery stock was being imported from the Orient and the South Pacific at that time, and because the pest closely resembled a scale pest of that country known as the dorthezia. Shortly after becoming Chief Entomologist for the federal government, Riley traveled to California to observe major pest problems there and took particular note again of the increasingly damaging scale pest. Because the cottony-cushion scale had first been described by a New Zealand entomologist, W. M. Maskell, Riley reiterated that the source of the invading pest must be the Austral-Asian region. This assumption was soon confirmed by Maskell, who in correspondence with W. G. Klee in San Francisco stated that cottony-cushion scale was indeed native to Australia.

In the mid 1880s, both Riley and Klee wrote about the possibilities of importing beneficial insects to control the cottony-cushion scale. Klee wrote the Australian entomologist Frazer Crawford to inquire about several possibilities. After some effort at procuring funds to support a natural enemy search, Riley, with the moral support of the California State Board of Horticulture, was able to assign U.S. Department of Agriculture entomologist Albert Koebele, then stationed in California and familiar with the scale pest, to this overseas venture. Koebele's mission took place in 1888.

Soon after arriving in Australia, Koebele found two enemies attacking the cottony-cushion scale on citrus: one a dipterous parasite, *Cryptochetum iceryae* (Williston), the other a coccinellid predator (lady bird beetle) commonly known as the vedalia, whose scientific name is *Rodolia cardinalis* (Mulsant). (Figure 3.1). Both were sent by ship to San Francisco, where they were examined, reared, and released in Los Angeles as adults on scale-infested citrus trees enclosed in canvas tents. The lady beetles immediately began feeding and ovipositing on scale infestations, rapidly increased in abundance, and quickly spread to adjacent trees.

The results of these colonizations were dramatic. The lady beetles multiplied and spread rapidly throughout the Southern California citrus-growing region. So did the other natural enemy, the cryptic little parasitic fly *Cryptochetum*, which had been colonized along with the vedalia. Scale infestations were reduced sharply, and within months the Southern California cottony-cushion scale epidemic had been reduced to harmless levels.

The Post-Vedalia Expansion of Biological Control

The marked success of the cottony-cushion scale project in California and its extension to many other parts of the world, coupled with its permanency, simplicity, and cheapness, led to enthusiastic support of similar ventures toward the solution of other agricultural pest problems. The method was envisioned as the utopian answer to age-old insect plague problems that have afflicted man throughout his history. Albert Koebele, who returned to the United States early in 1889, went back to Australia to look for more natural enemies of insect pests of concern to California. On this second mission (1892), Koebele concentrated mainly on predatory species, and the great majority of the enemies found and shipped back to California on this second trip were coccinellids. This was very likely because of the impressive success of the previous discovery, the vedalia beetle. Thus he sent to California *Cryptolaemus montrouzieri* (Mulsant), a lady beetle predator of another group of serious pests, the mealybugs, *Planococcus* spp. This predator, soon given the name mealybug destroyer, became well established for a time in Southern California, particularly along the coast, where it was a voracious attacker of citrus mealybugs. But any potential for continuous control was noted to be inhibited by its sensitivity to winter conditions each year. This maladaptedness led to reviving the idea of artificial propagation of the beetle in insectaries (an idea proposed in 1882 by Felix Gillet, an early California horticulturalist), with the release of large numbers in infested groves each spring and summer. This technique, now called *periodic colonization*, was put into effect in 1919 with considerable success.

One of the least effective beetle predators, *Rhizobius ventralis* Erichson,

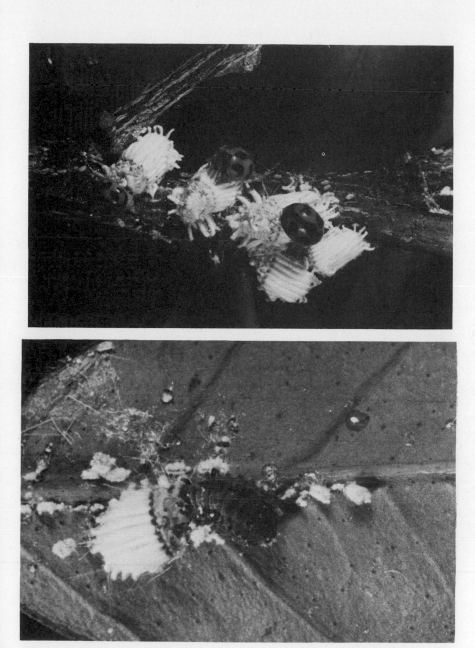

Figure 3.1. (A) Vedalia (*Rodolia cardinalis*) attacking the cottony-cushion scale, *Icerya purchasi*. (B) Closeup of a vedalia larva feeding on a scale (J. K. Clark).

introduced into California in 1892, holds historical interest in another way relative to biological control. Introduced as a predator of black scale, *Saissetia oleae,* it seems to have been involved in what may be the first reported observation of the *interference with natural-enemy action by pesticides.* In 1893 in Santa Barbara, where kerosene emulsion was used as a spray for controlling black scale on olive trees, it was noted that where sprays were used, scale infestations remained abundant, but no *R. ventralis* were to be found. On the other hand, in nearby unsprayed olive trees, the scale was observed to be less abundant, and *R. ventralis* was much more numerous. The conclusion then was that the emulsion must have interfered with the activity of the predator, possibly because of its repellent odor (Craw, 1894).

Twentieth-Century Developments

By 1900 the number of programs in biological control, the development of techniques for handling natural enemies, and the necessary facilities for such work had increased notably. Activity in California continued apace. In 1901 the parasite *Scutellista cyanea* Motschulsky, which a few years earlier had been introduced by the U.S. Department of Agriculture into Louisiana, was again imported into the United States, this time from South Africa to California for use against black scale. In 1903 the California State Horticultural Commission constructed an insectary in San Francisco, a special facility designed to receive and propagate imported natural enemies. This was the first such specialized facility in the nation for supporting biological control work. Destroyed by fire after the San Francisco earthquake of 1906, the state insectary was relocated to Sacramento in 1907.

In 1904 the horticultural commission initiated a project for the biological control of the codling moth, *Laspeyresia pomonella*, importing the parasite *Ephialtes caudatus* (Ratzeburg) from Spain. The campaigns against the black scale and the California red scale, initiated soon after the vedalia project, continued through the first decades of the new century.

Major projects were started elsewhere in the United States during this time, among them the cooperative one begun in 1905 between the state of Massachusetts and the U.S. Department of Agriculture against the gypsy moth, *Porthetria dispar*. This serious lepidopterous pest of numerous deciduous trees and shrubs in the northeastern United States apparently invaded America about 1869. A noteworthy feature of this program was the construction of a gypsy moth parasite laboratory designed to receive and process natural enemies of the pest as these were received from explorative work in Europe. At the same time a control program was started against the brown-tail moth, *Nygmia phaeorrhoea* (Donovan), in New England.

A pest of wheat in the midwestern United States became another focus of attention. This was an aphid, the greenbug, *Schizaphis graminum* (Rondani). A native parasite, *Aphidius testaceipes* (Cresson), heavily attacks the pest in the southern parts of its range but is much less effective in more northerly regions. University of Kansas scientists, in 1907, collected and distributed large numbers of dead parasitized aphids, known as mummies, into northerly fields in order to increase parasitism there. The results were variable, and claims of successful control of the greenbug as a consequence of this effort were confounded by other influences on the dynamics of the pest populations. A similar effort, aimed at collecting and distributing stocks of the native, aphid-feeding lady beetle *Hippodamia convergens* (Guerin) (Figure 3.2) collected from overwintering sites to various vegetable-growing areas of California, was begun in 1910 by the California State Horticultural Commission with indifferent result, and was later found to be ineffective because diapause beetles must feed, otherwise they fly away.

Elsewhere, biological control work increased from 1900 to 1910. As can be seen from Table 3.1, there were throughout the world at least eleven different programs of biological control, seven of which were judged to be completely or substantially successful. Countries involved included Australia, the United States (Hawaii), Italy, and Peru (DeBach, 1964). For a more recent review of biological control successes, see Laing and Hamai (1976).

Quarantine Considerations

It was soon realized that these beneficial insect importations could be quite hazardous unless done with great care by experts taking special precautions. The danger of accidental introduction of new insect pests, plant pathogens, or parasites of the natural enemies themselves during the routine handling of importation parcels suggested the need for strict quarantine security. The U.S. Department of Agriculture reserved to itself authority to permit shipments of insects and plants or plant parts from foreign sources into the United States. The first quarantine facility designed specifically for this purpose was constructed in Hawaii in 1913. It was an insect-proof laboratory with access limited to authorized personnel. Within the next several decades, all centers for the receipt of imported natural enemies were required by the U.S. Department of Agriculture to provide such facilities.

In 1913 in California, an experienced biological control specialist, Harry S. Smith, then employed at the gypsy moth parasite laboratory in Massachusetts, was appointed superintendent of the state insectary at Sacramento. Smith promptly increased the level of foreign exploration activity by California pest control workers, and he himself set out soon after appointment on his first overseas mission to Japan and the Philippines in search of enemies of the black scale. In 1914 Smith employed H. L. Viereck to search in southern

Figure 3.2. The ladybird beetle, *Hippodamia convergens*, an important naturally occuring predator of aphids in California (F. E. Skinner).

TABLE 3.1. The Progress of Biological Control Work, 1890–1975, by Decades[a]

Decade	Completely or substantially successful	Partially successful	Total
1890–1900	1	1	2
1900–1910	7	4	11
1910–1920	6	8	14
1920–1930	17	11	28
1930–1940	32	25	57
1940–1950	10	12	22
1950–1960	9	5	14
1960–1970	14	0	14+
1970–1975[b]	9	5	14

[a] Each project judged completely or substantially by DeBach (1964) is tabulated at the time of the initial establishment of the natural enemy involved.
[b] From Laing and Hamai (1976).

Europe for enemies of mealybugs. In Sicily, Viereck discovered and sent to Sacramento by ship the small encyrtid parasite *Leptomastidea abnormis* (Girault), a very effective enemy of the citrus mealybug. This parasite, first released in southern California in 1914, soon spread widely, resulting in a renewed interest in the biological control method and the consequent intensification of work in the state insectary under Smith.

During the decade 1910–1920, more than a dozen cases of establishment of natural enemies for control of pests were recorded around the world. At least six of these were judged to have been completely or substantially successful (see Table 3.1), involving the following pests:

Brown-tail moth	Canada
Anomala beetle	Hawaii
Alfalfa weevil	Utah
Rhinoceros beetle	Mauritius
Sugarcane weevil	Hawaii
Larch sawfly	Canada

In 1919 the U.S. Department of Agriculture created a laboratory in France to serve as a base of operations for the collection, rearing, and identification of natural enemies of the European corn borer, a major pest of corn and other crops in central North America. To reduce the possibilities of accidental introduction of harmful organisms to America, corn borer natural enemies were reared to the adult stage and shipped free of their hosts, of any of their own natural enemies (hyperparasites) that might be present, and of all plant material that might harbor, undetected, other insect and disease pests. This meticulous care to prevent the transfer to the United States of noxious species during natural-enemy importation work remains the hallmark of all biological control agencies. This laboratory is still functioning, working on such projects as cereal leaf beetle, *Oulema melanopus* L., green bug, *Schizaphis gramminum*, and shade tree and forest pests.

As can be seen from Table 3.1, biological control work intensified through the years 1920 to 1940. In 1923 the biological control work of the state of California, along with its leader, H. S. Smith, was transferred to the University of California Citrus Experiment Station, Riverside. A new quarantine–insectary facility was constructed there in 1929, and the old one in Sacramento was shut down. From this time on, as a consequence of Smith's enthusiastic and inspiring devotion to the study and promotion of biological control, responsibility in California for all biological control work lay with entomologists in university service, an arrangement until recently unique in California and the nation.

Smith's stimulating advocacy of biological control was not limited to California. His frequent contacts with pest control specialists in other nations inspired much additional biological control activity throughout the world. In the decade 1920–1930, more than 30 cases of natural enemy establishment were recorded. Among the important projects during this period was the dissemination of *Aphelinus mali* Haldeman, the parasite of the wooly apple aphid, *Eriosoma lanigerum* (Hausm.), from its native home in New England to New Zealand (where it soon provided very successful control of this apple pest), and then to British Columbia, Chile, South Africa, Italy, Uruguay, Brazil, Australia, and many other countries. Also noteworthy are the cases of control of the sugarcane leafhopper in Hawaii, the citrus blackfly in Cuba, the citrophilus mealybug in California, and the greenhouse whitefly, *Trialeurodes vaporariorum* Westw., in Canada, among others. In the United States, such major campaigns as the European corn borer project, the Japanese beetle program, and the oriental fruit moth project were started during this period.

The decade 1930–1940 saw a peak of activity throughout the world, with 57 different natural enemies established at various places. At least 32 of these led to successful control results. Then, as can be seen in Table 3.1 there followed a sharp drop in biological control activity, mainly because of World War II. However, this drop was also a consequence of the substantial reduction in the United States Department of Agriculture program of biological control, presumably because of a feeling that returns were incommensurate with the efforts expended. The worldwide discovery and subsequent emphasis upon use of synthetic organic insecticides beginning in 1945 more or less preempted the revival of biological control on any significant scale.

On the other hand, in California, biological control work continued unabated. In 1945 a new biological control laboratory was established by the University of California at its Gill Tract facilities in Albany, near Berkeley. The initial focus of this new laboratory was on control of the oriental fruit moth and the St. Johnswort or Klamath weed. The latter project was highly successful (see Chapter 9).

During the 1950s, such pests as the spotted alfalfa aphid, *Therioaphis trifolii* (Monell); pea aphid, *Acyrthosiphon pisum* (Harris); and olive scale, *Parlatoria oleae* (Clovée), were successfully controlled. During the next decade, California red scale, *Aonidiella aurantii*; puncture vine, *Tribulus terrestris* L. (Figure 3.3); and the Egyptian alfalfa weevil, *Hypera brunneipennis*, were studied. Only the latter pest has not been controlled, except by newly developed winter insecticide treatments designed to minimize disruption of biological control of other pests. The decade 1970–1980 provided some dramatic examples of biological control, among them walnut aphid, *Chromaphis juglandicola* Kalt., and linden aphid, *Eucallipterus tilliae* L. Other pests such as the dusky veined aphid, *Callaphis juglandis*; iceplant scale, *Pulvinaria mesembryanthemi* (Vallott); pink bollworm, *Pectinophora gossypiella* (Saunders); bindweed, *Convovulus arvensis* L.; and the blue alfalfa aphid, *Acyrthosiphon kondoi* (Shinji), are currently being investigated.

In other areas of the country, such pests as alligator weed, *Alternanthera philoxeroides* (Mart.) Grisch; cereal leaf beetle, *Oulema melanopus*; alfalfa weevil, *Hypera postica* Gyll.; hydrilla weed, *Hydrilla verticillata* (L.f.) Royle; and water hyacynth, *Eichornia crassipes* (Mart.) Solms-Laubach, were partly to completely controlled. Other long-standing pests such as codling moth, *Laspeyresia pomonella*, and gypsy moth, *Porthetria dispar*, are still subjects of continuing biological control with only marginal success.

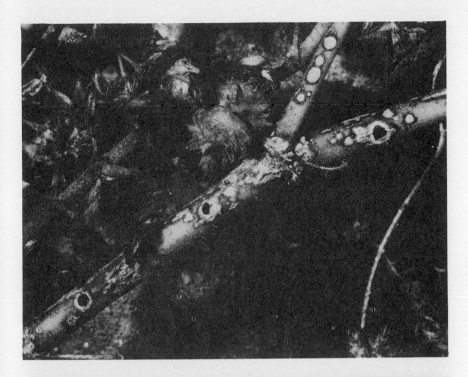

Figure 3.3. Puncturevine, *Tribulus terrestris*. (A) Stems of puncturevine showing damage to the stem by the weevil, *Microlarinus lypriformis*. (B) *Microlarinus lareynii* adult on the spiny seed of the puncturevine. Note the hole left by the emerging weevil in the seed (J. K. Clark).

Development of International Organizations

By the mid-1920s, many of the dominions and colonies of the British Empire were active in biological control work, including Australia, New Zealand, Fiji, Canada, Bermuda, and South Africa. In some of these countries, biological control facilities and teams were permanently employed. At Belleville, Ontario, the Canada Department of Agriculture constructed a biological control laboratory in 1929, and in 1936 a quarantine reception center was added. However, in 1971 this facility was closed and the quarantine reception activity transferred to Ottawa.

In 1927 the Imperial Bureau of Entomology created a special facility for the conduct of biological control work in England, the Farnham House Laboratory. This institution, which in 1928 came under the direction of W. R. Thompson, accepted requests for study, exploration, and delivery of natural enemies for use against specified pests of importance to the various countries of the British Empire. Concentrating first on enemies of insect pests, in 1929 it broadened its scope to include insect enemies of weeds. At Farnham House Laboratory a variety of services were provided, including a bibliographic service pertaining to literature on biological control, the cataloging of important pests and their natural enemies, the preliminary surveying by field investigators of pest-infested areas for additional natural enemies and for ecological information about their distribution and abundance, the collection of living samples of these enemies and their identification, the determination of their life histories and requirements for culture, and the development of appropriate shipping techniques for transferring selected natural enemies to various recipient agencies throughout the world.

Early projects undertaken at Farnham House Laboratory included the search for the study of enemies of the codling moth, the pear slug, *Caliroa cerasi* (L.); the European fruit lecanium, *Eulecanium coryli* (L.); the oriental fruit moth, *Grapholitha molesta* (Busck); the pine shoot moth, *Rhyacionia buoliana* Achiffermuller; sawflies of the genus *Sirex*; the larch case bearer, *Coleophora laricella* Hubner; the lucerne flea, *Sminthurus viridis* Lbb; the wheat-stem sawfly, *Cephus cinctus* Norton; the diamondback moth, *Plutella maculipennis* Curtis; the carrot rust fly, *Psila rosa* L.; the pink bollworm, *Pectinophora gossypiella*; the greenhouse whitefly, *Trialeurodes vaporariorum* (Westw.); the sheep blowfly, *Lucilia sericata* Meigen; and the horn fly, *Lyperosia irritans* L. Stocks of the parasite *Aphelinus mali*, derived from the United States, were maintained for distribution to requesters for the control of the woolly apple aphid.

In 1940 the Empire biological control facility was moved because of the war to Ottawa, Canada, where it became known as the Imperial Parasite Service, still a component of the Imperial Institute of Entomology. In 1947 the service became independent and was designated as the Commonwealth Bureau of Biological Control. A few years later in 1951, the facility received its present name, Commonwealth Institute for Biological Control, or CIBC. In

1961 the CIBC headquarters was transferred to Trinidad, West Indies, where it remains today.

Starting with but one laboratory–office–insectary facility at Farnham Royal, United Kingdom, the CIBC embraced at its peak not only an administrative headquarters and laboratory but also regional stations and substations in Argentina, Switzerland, India, Malaysia, Pakistan, Uganda, West Africa, and the West Indies. With financial support from the Commonwealth countries, the institute carries out, by request, foreign exploration for natural enemies, studies on life histories and ecologies of various pests and their natural enemies in their areas of indigeneity, and the forwarding of shipments of natural enemies.

The CIBC was the first truly worldwide biological control organization. Besides providing services to Commonwealth countries, it also responds to requests from other nations at cost. The U.S. Department of Agriculture and the University of California have at various times received information and parasite materials from the CIBC and vice versa. In 1970 the program of the CIBC included research work centered at some 23 stations and substations involving 118 different projects.

In 1948, under the auspices of the International Union of Biological Sciences, an international conference was held at Stockholm, Sweden, composed of biological experts and administrators from many countries of the world, for the purpose of considering the establishment of an international organization for biological control. A country or individual joining such an organization would be entitled to receive information and services about natural enemies of pest species and would have the opportunity to acquire, at cost, shipments of beneficial organisms from other members of member organizations.

Because many countries and institutions in attendance at the meeting already provided for their own biological control services, including facilities and staffs of experts capable of supplying much of the expertise needed to carry out biological control programs, most of the institutions and governments, including the U. S. Department of Agriculture, California, Hawaii, the British Commonwealth, many Commonwealth countries, and the Soviet Union, decided not to join such an international organization. As a result, when, in 1955, the Commission Internationale de Lutte Biologique contre les Enemis des Cultures came into being, it principally included countries and member institutions from European, Mediterranean, and Near Eastern regions. The CILB established headquarters at Zurich, sponsored a quarterly scientific journal, *Entomophaga*, and maintained a taxonomic service in Geneva and a bibliographic service centered at Darmstadt, Germany. Working groups composed of experts appointed by the general secretariat met on occasion to consider and advise on possible solutions of particular problems, such as the biological control of the Colorado potato beetle in Europe.

In 1962 the CILB undertook a reorganization, changing its name to the Organisation Internationale de Lutte Biologique contre les Animaux et les

Plants Nuisibles. In 1971, again through the stimulus of the International Union of Biological Sciences, the OILB was further revamped into a true worldwide organization composed of regional sections and a parent global organization. This organization has now been widely accepted as the definitive organization of biological control specialists on a worldwide basis. The success of OILB is most important now, as considerable interest in biological control is developing in many underdeveloped countries. In South America, a large regional project on the biological control of the pea aphid, *Acyrthosiphon pisum*; blue aphid, *A. Kondoi*; and several cereal aphids, *Metopolophium dirhodum* Walker, *Sitobion avenae* F., *Rhopalosiphum maidis* Fitch, and *R. padi* L., have been moderately to highly successful. These cereal pests in conjunction with plant diseases reduce yields by more than 80% in Brazil alone. The potential for biological control in Latin America is exceedingly great (Hagen, 1977). In general, biological control in underdeveloped countries has barely scratched the surface.

References

Clausen, C. P. 1940. *Entomophagous Insects*. McGraw-Hill: New York. 688 pp.

Clausen, C. P. 1956. Biological control of insect pests in the continental United States. U. S. Dept. Agric. Tech. Bull. 1139, 151 pp

Craw, A. 1894. Biennial report of quarantine officer and entomologist. Fourth Bienn. Rep. St. Bd. of Hort. St. of Calif. Sacramento. 1893–1894.

DeBach, P. (ed.) 1964. *Biological Control of Insect Pests and Weeds*. Chapman & Hall: London. 844 pp.

DeBach, P. 1974. *Biological Control by Natural Enemies*. Cambridge University Press: London. 323 pp.

Doutt, R. L. 1964. The historical development of biological control. In: P. DeBach (ed.) *Biological Control of Insect Pests and Weeds*, Chap. 2. Chapman & Hall: London. pp. 21–42.

Essig, E. O. 1931. *A History of Entomology*. Macmillan: New York. 1029 pp.

Gillet, F. 1882. First report. Calif. St. Hort. Comm., Sacramento. pp. 24–28

Hagen, K. S. 1977. Biological insect control. Insect pests of Brazil: Graduate training, accomplishments and future. Mich. St. Univ. Brazil Project M.E.C. Rept. #55.

Hagen, K. S. and J. M. Franz. 1973. A history of biological control. In: R. F. Smith, T. E. Mittler, and C. N. Smith (eds.) *History of Entomology*. Annual Reviews: Palo Alto. pp. 433–477.

Huffaker, C. B., and P. S. Messenger (eds.) 1976. *Theory and Practice of Biological Control*. Academic Press: New York. p. 788.

Laing, J. E., and J. Hamai. 1976. Biological control of insect pests and weeds by imported parasites, predators and pathogens. In: C. B. Huffaker and P. S. Messenger (eds.) *Theory and Practice of Biological Control*. Academic Press: New York. pp. 685–693.

McCook, H. 1882. Ants as beneficial insecticides. *Proc. Acad. Nat. Sci.*, Philadelphia.

Swan, L.A. 1964. *Beneficial Insects*. Harper & Row: New York. 429 pp.

Sweetman, H. L. 1936. *The Biological Control of Insects*. Comstock Publishing Associates: Ithaca, New York. 461 pp.

Sweetman, L. A. 1958. *The Principles of Biological Control*. W. C. Brown: Dubuque. 560 pp.

4

Natural Enemies

The term *entomophagy* is a combination of the Greek words *entomon* (insect) and *phagein* (to eat), and thus denotes the insect-eating habit. There is also an adjectival form of this Greek word combination, as when we speak of *entomophagous* insects. Finally, the same Greek combination is used in a generic sense when we lump the insect-eaters together as *entomophaga*.

In this volume our interest is largely with entomophagous insects. However, other kinds of animals also feed upon insects and often play important roles in insect population regulation. The Nematoda include some important parasites of insects. Among the noninsectan arthropods there is a broad spectrum of insect-feeders, including scorpions, spiders, mites, sunspiders, pseudoscorpions, and harvestmen. In the higher groups there are entomophagous amphibians, fish, reptiles, birds, marsupials, and mammals, some of which are exclusively insectivorous. Even in man, the degree of entomophagy is surprisingly well developed (Bodenheimer, 1951). Among the California Indians, oak wax scale, grasshoppers, wood-boring beetle larvae, June beetle adults, crane fly larvae, fly puparia, adult flies, salmon fly nymphs and adults, a variety of caterpillars, and bee and yellow jacket larvae were all utilized as food (Essig, 1931).

Insect feeding in birds of various species has been well studied, and a number of species have been found to be of considerable importance in controlling forest insects (Buckner, 1966, 1967, 1971; Coppell and Sloan, 1970; and Dahlsten and Copper, 1979) and the codling moth (MacClelland, 1970). Shrews and mice are also important enemies of forest insects (see Holling, 1966). Various fishes and amphibians are important in the biological control of aquatic insects, some of which (e.g., mosquitoes, gnats, midges) are pests (Gerberich, 1946; Gerberich and Laird, 1968; Sweetman, 1958; Bay *et al.*, 1976).

In applied biological control, several nonarthropod species have been deliberately used with some measure of success. Included are *Gambusia affinis* (Baird and Girard) and other fishes against mosquitoes, gnats, and midges (Gerberich, 1946; Gerberich and Laird, 1968; Bay *et al.*, 1976); the nematode *Neoaplectana glaseri* Steiner against the Japanese beetle; the giant toad *Bufo marinus* L. for control of white grubs and sugarcane rhinoceros beetle (Sweetman, 1958); and the mynah bird, *Acridotheres tristis* (L.), against the red locust (Sweetman, 1958). But the more varied the repertoire of behavior of vertebrates, particularly learning and propensity for switching prey, the more difficult it is to predict their host range and thus the possible consequences of their introduction for biological control. For example, the introduction of the mongoose in the Hawaiian Islands for rat control resulted in the extinction of some ground nesting birds; while in sugarcane in Australia, the giant toad became so numerous that it has become a nuisance in its own right and more recently has hampered the biological control of dung by dung beetles. These problems merely stress the point that great care matured in long years of experience is required in biological control—it is not work for amateurs.

Despite the wide spectrum of insect-eating organisms, the role of the noninsect predators, particularly in classical biological control, has been minor, which is why the emphasis of this volume is on the use of insects in biological control.

Entomophagous insects fall into two categories, *predators* and *parasites* (parasitoids). The two groups differ in several ways. Characteristically, a predator is relatively large compared to its host (prey), which it seizes and either devours or sucks dry of its body fluids rather quickly. Typically, the individual predator consumes a number of prey (e.g., a single lady beetle larva may consume hundreds of aphids) in completing its development. Most often a predator is carnivorous in both its immature and adult stages and feeds on the same kind of prey in both stages, but there are exceptions (e.g., syrphid flies and green lacewings).

By contrast, the parasitoid is invariably parasitic only in its immature stages and develops within or upon a single host, that is slowly destroyed as the parasitic larva completes its development. The parasitoid adult is usually free living, feeding on such foodstuffs as nectar, honeydue, and sometimes host insect body fluids. The biologies of parasitic species are immensely varied and fascinating.

Predators

Predatory insects are the lions, wolves, sharks, barracudas, hawks, and shrikes of the insect world (see Figure 4.1). The predatory insect either lies in wait to pounce upon its unsuspecting victim, runs it down, or—where the host is sessile or semisessile—may literally browse off the population.

Predatory insects feed on all host stages: egg, larval (or nymphal), pupal,

Figure 4.1. An assortment of insect and vertebrate predators.

Figure 4.1A. Mosquito fish, *Gambusia affinis* (P. F. Daley).

Figure 4.1B. An assassin bug nymph, *Zelus renardii* Kol. (K. Middleham).

Figure 4.1C. An adult green lacewing, *Chrysopa carnea* (P. F. Daley).

Figure 4.1D. Larva of *C. carnea* feeding on a pear psylla (J. K. Clark).

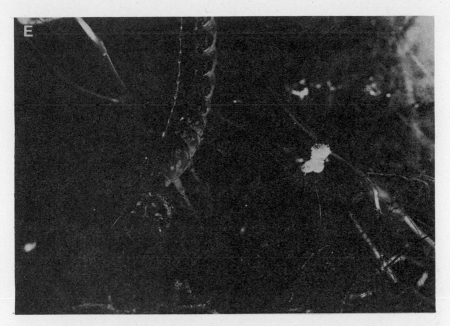

Figure 4.1E. Larva of an aquatic diving beetle, *Dytiscus* sp. (P. F. Daley).

Figure 4.1F. A praying mantis (K. Middleham).

Figure 4.1G. A mantispid (P. F. Daley).

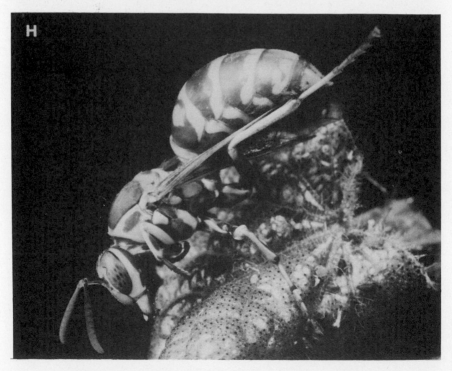

Figure 4.1H. A predaceous wasp, *Polistes apachus* (K. Middleham).

Figure 4.1I. Adult ladybird beetle, *Hippodamia convergens* (J. K. Clark).

Figure 4.1J. Larva of a ladybird beetle, *Coccinella* sp. (F. E. Skinner).

Figure 4.1K. An adult syrphid (F. E. Skinner).

Figure 4.1L. A larval syrphid (F. E. Skinner).

and adult. From the standpoint of feeding habit, there are two kinds of predators: those with chewing mouth parts, e.g., lady beetles (Coccinellidae) and ground beetles (Carabidae), which simply chew up and bolt down their victims—legs, bristles, antennae, and all—and those with piercing mouthparts, e.g., assassin bugs (Reduviidae), lacewing larvae (Chrysopidae), and hover fly larvae (Syrphidae), which suck the juices from their victims. The sucking type of feeder often injects powerful toxins and digestive enzymes that quickly immobilize the prey so that the feeding process is a placid affair with little thrashing about by the victim. For example, once a lacewing larva clamps its sicklelike mandibles into a caterpillar several times its size, the latter is doomed, and its period of struggle lasts but a few seconds. The digestive substances injected into prey facilitate the intake of body contents.

Predatory species occur in most insect orders, with the greatest number of species occurring in the Coleoptera. One order, the Odonata (dragonflies) is exclusively predaceous, and others are nearly so. Predators may be *poly-phagous*, having a broad host range (e.g., the green lacewing, *Chrysopa carnea* Stephens); *oligophagous*, having a restricted host range (e.g., aphid-feeding coccinellids and syrphids); or essentially *monophagous*, that is, highly prey specific (e.g., *Rodolia cardinalis*, which feeds only on cottony-cushion scale, *Icerya purchasi*, and its close relatives). With *R. cardinalis* the parasitic habit is approached, since the female beetle deposits its egg on an adult female scale or on a single scale's egg mass, and the hatching larva can complete its development on the eggs supplied by this initial source (Clausen, 1940). A number of predatory insects resemble parasites in this way, and there are a number of parasitoids whose habits approach those of predators. Many entomologists, in fact, consider the parasitoids to be specialized predators rather than parasites.

More recent studies have shown that some predators such as *Geocoris pallens* Stål also suck juices from nonvital plant parts, while some herbivores such as *Lygus hesperus* are also insect predators. These omnivorous habits might be more widespread than formerly suspected. In such predators, meat is necessary for reproduction and growth, while plant sugars are used to meet maintenance respiration requirements.

Although predators have been far overshadowed by parasites in classical biological control, they nevertheless have been of great significance in several programs—e.g., *R. cardinalis* versus cottony-cushion scale and *Tytthus* (=*Cyrtorrhinus*) *mundulus* (Breddin) versus sugarcane leafhopper—and their importance in naturally occurring biological control is inestimable. For example, in California's San Joaquin Valley, predators are thought to be more important in restraining the major lepidopterous pests (i.e., bollworm, cabbage looper, beet armyworm) then are parasites. Soybean in the southern U. S. is another crop wherein predators are considered to be more important. The worldwide eruption of spider mites in the wake of widespread use of chemical insecticides has mainly resulted from the elimination of predators of spider mites by the insecticides (McMurtry *et al.*, 1970; Huffaker *et al.*, 1970). Attempts have more recently been made to introduce pesticide-resistant mite predators into apple orchards and grape vineyards to overcome the disruption caused by pesticide applications made to control herbivorous mites (Croft, 1977; Hoy and Knop, 1981).

Parasites

An insect that parasitizes other insects is known as a parasitoid, a term that distinguishes these entomophagous insects from all other kinds of parasites (see Figure 4.2). No other group of organisms parasitizes its own kind to the extent that insects do. In fact, in insects this habit has reached the remarkable extreme of *adelphoparasitism*, wherein a species is parasitic on itself, as with certain Aphelinidae in which males are obligate parasites of females of their own species (e.g., *Coccophagus scutellaris* Dalman, *Coccophagoides utilis* Doutt).

Parasitoids are recorded from five insect orders, with the bulk of the species occurring in the Diptera and Hymenoptera (also Coleoptera, Lepidoptera, and Strepsiptera). Despite their restriction to only five orders, there are tremendous numbers of parasitoid species worldwide. Kerrich (1960), extrapolating from the numbers of described Coleoptera and parasitic Hymenoptera in the well-studied British fauna (4000 beetle species and 5000 species of parasitic Hymenoptera), estimates that up to 500,000 parasitic Hymenoptera might be described worldwide if that fauna were to be as well studied as that of beetles (of which, by his estimates, there were 300,000 described species in 1960).

Kerrich also stated that W.H. Ashmead's 60 year-old guess of a million ichneumonids alone "still seems sensible today." Townes (1969) was considerably more conservative in his estimate of the number of ichneumonids, giving a figure of approximately 60,000 species—which is still a tremendous number. Although these numbers are only guesses, they are the projections of experienced scientists, and even if they are an order or two in magnitude off the mark, they still reflect the enormous abundance and variety of parasitiods.

Parasitoids attack and develop in all insect stages—egg, larval (nymphal), pupal, and adult—and again, as with the predators, the host ranges of individual species run the full spectrum from monophagy to polyphagy.

For example, the tachinid *Compsilura concinnata* Meigen has been recorded from more than 100 hosts representing 3 orders and 18 families (Clausen, 1940), while the aphidiid *Trioxys complanatus* Quilis is stenophagous, developing only in species of *Therioaphis*, and its close relative

Figure 4.2. A variety of insect parasites (parasitoids) (A–B, P. F. Daley, C–F, F. E. Skinner).

Figure 4.2A. *Praon* sp., a parasite of the dusky winged aphid, *Callaphis juglandis*.

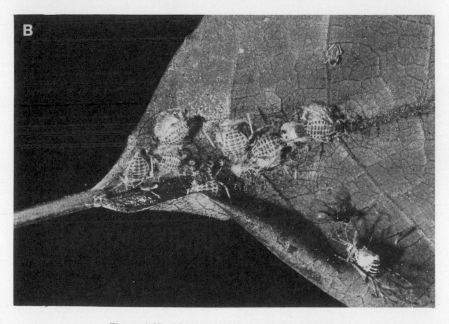

Figure 4.2B. Mummies of *Praon* in *C. juglandis*.

Figure 4.2C. A gregarious endoparasite, *Pentalitomastix plethoricus* Caltagirone, (pupae) in the navel orangeworm.

Figure 4.2D. Adult *P. plethoricus* from a single larva.

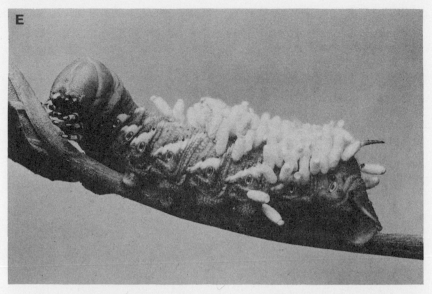

Figure 4.2E. *Apanteles congregatus* (Say) pupae on a hornworm larva. This parasite pupates externally. Note that many have emerged.

Figure 4.2F. Mummies of *Aphidius smithi* in the pea aphid.

Trioxys pallidus (Haliday) is apparently monophagous on *Chromaphis juglandicola* (Kalt.).

Kinds of Parasitoids

Insect parasitism manifests itself in a number of ways, and as a result, there is a considerable terminology that describes the kinds of parasites and the nature of their development.

Primary Parasite. Primary parasites are those species that develop in or upon nonparasitic hosts. These hosts may be phytophagous, saprophagous, coprophilous, polleniferous, fungiferous, or predatory, but in no case are they themselves parasitoids.

Hyperparasite. A hyperparasite is a parasitoid that develops on another parasitoid (i.e., a parasite of a parasite). There may be more than one level of hyperparasitism in a given relationship (Figure 4.3). Thus, in the relationship between the pea aphid and *Aphidius smithi* Sharma and Subba Rao, the primary parasite *A. smithi* may first be attacked by *Alloxysta victrix* (Westwood), which for precision may be called a secondary parasite (Gutierrez and van den Bosch, 1970), and *Alloxysta* in turn may be attacked by *Asaphes californicus* Girault, a tertiary parasite (Sullivan, 1969). *Asaphes* may also directly attack *Aphidius*, in which case it acts as a secondary parasite. Both *Alloxysta* and *Asaphes* are *direct hyperparasites* since they oviposit directly into or upon the primary species. But the two differ in their method of attack. *Alloxysta* attacks its host (*A. smithi*) when the latter is still contained with the living aphid. In doing this, an *Alloxysta* female penetrates the aphid integument with its ovipositor, seeks out the *Aphidius* larva with this organ, and deposits an egg internally in its victim. The hatching *Alloxysta* larva then develops as a solitary internal parasite of *Aphidius* and ultimately kills it in the prepupal stage. *Asaphes* attacks its hosts by drilling through the aphid mummy (parchmentlike skin of the dead pea aphid), venomizing the contained *Aphidius* or *Alloxysta*, and then depositing an egg on the surface of the victim so that the hatching larva develops as an external parasite. As mentioned above, species such as *Asaphes* and *Alloxysta* are termed direct hyperparasites because of their direct oviposition into or upon the victim. By contrast, some hyperparasites simply oviposit into an insect whether it is parasitized or not. The hatching larva then seeks out the primary parasite if it is present, or if not, waits until a primary parasite egg is deposited in the insect and attacks the hatching larva. If no primary parasite egg is deposited, the hyperparasite simply perishes for lack of a host. Species of this type of habit are termed *indirect hyperparasites*. An example of this type of hyperparasitism would be the oviposition of *Mesochoris* sp. eggs into unparasitized lepidopterous larvae in anticipation of parasitism by the primary parasites of the genus *Apanteles sp*.

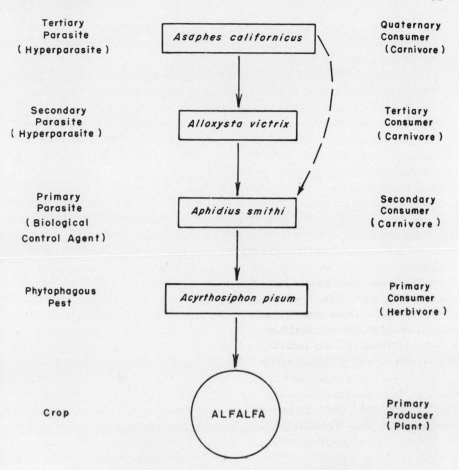

Figure 4.3. The biology of various hyperparasites of *Aphidius smithi*.

Endoparasitism. Parasitoids that develop within the host's body (internally) are called *endoparasites*. Where only a single larva completes its development in a given host, the species is termed a *solitary endoparasite*. Where several to many larvae develop to maturity in a single host, the term *gregarious endoparasite* is used.

Ectoparasitism. A species that develops externally (the larva feeds by inserting its mouthparts through the victim's integument) is known as an *ectoparasite*. Again, as with endoparasites, there are solitary and gregarious ectoparasites.

Multiple parasitism. A situation in which more than one parasitoid species occurs simultaneously within or upon a single host is termed *multiple parasitism*. In most cases only one of these species survives to maturity, the others succumbing to competitive interaction (fighting, physiological suppression). In rare cases, as with species of *Trichogramma* that parasitize insect eggs (particularly those of Lepidoptera), more than one species may complete its development in the egg.

Superparasitism. The phenomenon in which more individuals of a given parasitoid species occur in a host individual than can develop to maturity in that host is called *superparasitism*. Where this occurs with solitary endoparasites, internecine battle or physiological suppression of the supernumerary larvae or eggs results in the survival of a dominant individual. In some cases, however, the host itself succumbs prematurely before the supernumerary parasites are eliminated, and all perish.

Adelphoparasitism (autoparasitism). The phenomenon in which a species of parasitoid is parasitic upon itself, is termed *adelphoparasitism*. This is the case with *Coccophagus scutellaris*, the male of which is an obligate parasite of the female.

Cleptoparasitism. The phenomenon in which a parasitoid preferentially attacks hosts that are already parasitized by another species is called *cleptoparasitism*. The cleptoparasite is not hyperparasitic, for it does not parasitize the previously occurring parasite species. Instead, multiple parasitism is involved, and the relationship between the two species is competitive, with the cleptoparasite usually dominating.

Modes of Reproduction

In the parasitic Hymenoptera, several different patterns of reproduction occur that have important bearing on the ecology and habits of different species. These are all variations on a basic phenomenon in all the Hymenoptera known as *haploid parthenogenesis*. Haploid parthenogenesis refers to the fact that the unfertilized egg can undergo parthenogenetic development to produce a normal, viable adult. In every case, the haploid individual is a male. On the other hand, the fertilized egg develops into a diploid female adult. However, there are differences in the way some parasitoids follow this basic pattern. Such differences in mode of reproduction are termed *arrhenotoky, deuterotoky,* and *thelyotoky*.

Arrhenotoky. The basic reproductive mode is *arrhenotoky*, in which unfertilized eggs produce males and fertilized eggs produce females. Hence, virgin females can produce progeny, but they will be all male. Species that follow this mode of reproduction are referred to, for obvious reasons, as *biparental*. It is important to note that in some biparental species the mated female can produce either male or female offspring through external or internal control of fertilization. In other species the mated female produces only female progeny.

Deuterotoky. The reproductive mode by which unmated females produce

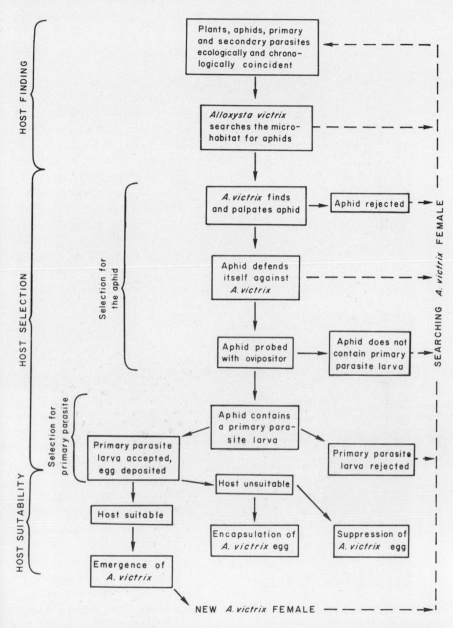

Figure 4.4. Host selection behavior of the direct hyperparasite *Alloxysta victrix* (adapted from Gutierrez, 1973).

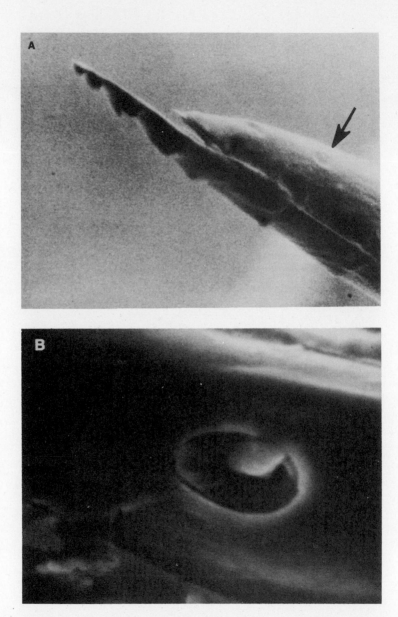

Figure 4.5. Sensory structures. (A) Ovipositor of *Alloxysta victrix* showing sensory pits (as at arrow) that enable it to distinguish among the primary parasite larvae of aphids (× 1500). (B) Same structure × 12,000 (photos by A. P. Gutierrez). (C) Sensory structures for the antenna of a larval lacewing (*Chrysopa* sp.) (× 3000, reproduced at 68%) (J. DeBenedictis).

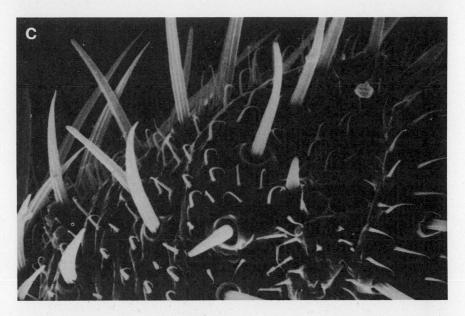

Figure 4.5. (*continued*)

both male and female progeny is termed *deuterotoky*. Such species are called *uniparental*. The haploid males produced in such cases are biologically and ecologically nonfunctional. The females produce in their female progeny a diploid condition through various cytogenetic mechanisms. Often in deu- terotokous species, the proportion of males produced relative to total progeny varies with external conditions such as temperature.

Thelyotoky. In thelyotoky, the virgin females produce only female pro- geny, and males are unknown. Such uniparental species are not uncommon. Similar cytogentic mechanisms for restoring the diploid condition in the egg occur here as in the case of deuterotoky. In a few cases, thelyotokous species, under stressful conditions of extreme temperature, for example, shift to the deuterotokous mode and produce haploid males as well as diploid females.

Host Specificity

We have mentioned host specificity, but there are several physiological and ecological factors that help maintain specificity in nature. First there are the broad categories of host finding, host selection, and *host suitability*, within which are imbedded more subtle behavior. Figure 4.4 shows the behavior sequence used by the hyperparasite *Alloxysta victrix* in selecting its hosts. This complicated biology represents an extreme, but all species employ some subset of these steps, using chemosensory structures on the ovipositor, antennae, mouth parts, tarsi, or other structures as aids to make many of the decisions at behavioral branch points (Figure 4.5).

Conclusion

It is apparent from the foregoing discussion that entomophagous insects are varied and abundant, and that they frequently develop complex interrelationships with their hosts and among themselves. Figure 4.6 depicts the intricate relationship of several insect pests and their natural enemies in California alfalfa. As incomplete as it is, the diagram strikingly illustrates the wide spectrum of entomophagous arthropods that can impinge upon single

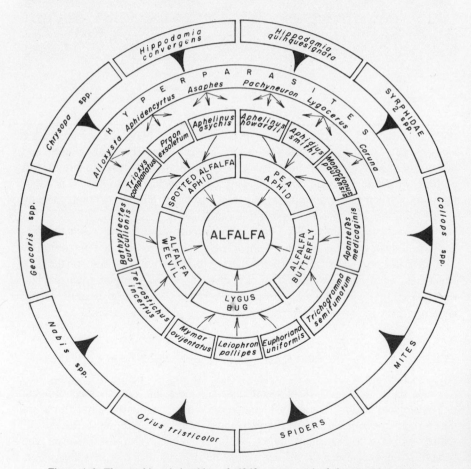

Figure 4.6. The trophic relationships of alfalfa, pests, and of the natural enemies.

phytophagous species in a simple agroecosystem. If we extrapolate from this model to the estimated 10,000+ pest insect species worldwide and consider the vast numbers of potentially injurious ones permanently restrained by natural enemies, the phenomenon of entomophagy assumes enormous significance because of both the diversity of predatory and parasitic species involved and the impact that they have on insects and insectlike arthropods.

References

Bay, E. C., C. O. Berg, H. C. Chapman, and E. F. Legner. 1976. Biological control of medical and veterinary pests. In: C. B. Huffaker and P. S. Messenger (eds.) *Theory and Practice of Biological Control*. Academic Press: New York. pp. 45–479.

Bodenheimer, F. S. 1951. *Insects as Human Food*. W. Junk: The Hague. 352 pp.

Buckner, C. H. 1966. The role of vertebrate predators in biological control of forest insects. *Annu. Rev. Entomol.* **11**:449–470.

Buckner, C. H. 1967. Avian and mammalian predators of forest insects. *Entomophaga* **12**:491–501

Buckner, C. H. 1971. Vertebrate predators. In: *Toward Integrated Control*. Proc. Third Ann. N.E. For. Ins. Work Conf. USDA For. Serv. Res. Paper NE-194. pp. 21–31.

Clausen, C. P. 1940. *Entomophagous Insects*. McGraw-Hill: New York. 688 pp.

Coppel, H. C., and N. F. Sloan. 1970. Avian predation, an important adjunct in the supression of larch casebearer and introduced pine sawfly populations in Wisconsin forests. *Proc. Tall Timbers Conf. Ecol. Anim Contr. Habitat Manage*. **2**:259–272.

Croft, B. A. 1977. Resistance in arthropod predators and parasites. In: D. L. W. Watson and A. W. A. Brown (eds.) *Pesticide Management and Resistance*. Academic Press: New York.

Dahlsten, D. L., and W. A. Copper 1979. The use of nesting boxes to study the biology of the mountain chickadee *(Parus gambeli)* and its impact on selected forest insects. In: J. G. Jackson, R. N. Conner, R. R. Fleet, J. A. Jackson, and C. R. Knoll (eds.) *The Role of Insectivorous Birds in Forest Ecosystems*. Academic Press: New York. pp. 217–259.

Essig, E. O. 1931. *A History of Entomology*. Macmillan: New York. 1092 pp.

Gerberich, J. P. 1946. An annotated bibliography of papers relating to the control of mosquitoes by the use of fish. *Am. Midl. Nat.* **36**:87–131.

Gerberich, J. B., and M. Laird. 1968. Bibliography of papers relating to the control of mosquitoes by the use of fish. FAO Fisheries Tech. Paper No. 75 (FRs/T75). 70 pp.

Gutierrez, A. P. 1973. Studies on host selection and host specificity in the aphid hyperparasite *Charips victrix* (Hymenoptera: Cynipidae). 6. Description of sensory structures and a synopsis of host selection and host specificity. *Ann. Entomol. Soc. Am.* **63**:1705–1709.

Gutierrez, A. P., and R. van den Bosch. 1970. Studies on host selection and host specificity in the aphid hyperparasite *Charips victrix* (Hymenoptera: Cynipidae). 1. Review of hyperparasitism and the field ecology of *Charips victrix*. *Ann. Entomol. Soc. Am.* **63**:1345–1354.

Holling, C. S. 1966. The functional response of invertebrate predators to prey density. *Mem. Entomol. Soc. Can.* **48**:3–86.

Hoy, M. A., and N. F. Knop. 1981. Selection for and genetic analysis of permethrin resistance in *Metaseiulus occidentalis*; genetic improvement of a biological control agent. *Entomol. Exp. Appl.* **28**:(in press).

Huffaker, C. V., M. van de Vrie, and J. A. McMurtry. 1970. Ecology of tetranychid mites and their natural enemies: A review. II. Tetranychid populations and their possible control by predators: An evaluation. *Hilgardia* **40**:391–458.

Kerrich, G. J. 1960. The state of our knowledge of the systematics of the Hymenoptera parasitica. *Trans. Soc. Brit. Entomol.* **14**:1–18.

MacClelland, C. R. 1970. Woodpecker ecology in the apple orchard environment. *Proc. Tall Timbers Conf. Ecol. on Anim. Contr. Habitat Manage.* **2**:273–284.

McMurtry, J. A., C. B. Huffaker, and M. Van de Vrie. 1970. Ecology of tetranychid mites and their natural enemies: A review. I. Tetranychid enemies. Their biological characters and the impact of spray practices. *Hilgardia* **40**:331–390.

Sullivan, D. J. 1969. Study of aphid hyperparasitism with special references to *Asaphes californicus* Girault (Hymenoptera: Pteromalidae). Ph.D. Thesis. University of California, Berkeley.

Sweetman, L. A. 1958. *The Principles of Biological Control.* W. C. Brown: Dubuque. 560 pp.

Townes, H. 1969. The genera of Ichneumonidae. Part 1. *Mem. Am. Entomol. Inst.* No. 11. 300 pp.

5

Microbial Control of Insects, Weeds, and Plant Pathogens

The Definition and Scope of Microbial Control

Microbial control is here defined as "The utilization of pathogens for the management of pest populations." Nematodes, although not microorganisms, are considered to be "microbial control" agents because of the techniques involved in their utilization. Various types of microorganisms have been used in biological control, among them bacteria, viruses, fungi, and protozoa. The most important groups of microbial agents are discussed in this chapter. In the broadest consideration, pathogens are important naturally occurring biological control agents. In nature, microbial pathogens frequently cause epiphytotics and epizootics, which help maintain the balance of plant and animal populations. The role of microbes in the dynamics of organisms was observed early on by biologists, who eventually initiated studies to attempt to manipulate pathogens for pest control. This was the genesis of the microbial control tactic that in recent years has become a more significant weapon in the pest control arsenal. Microbial control has predominantly involved the artificial manipulation of pathogens, especially viruses, bacteria, fungi, protozoans, and nematodes formulated as sprays or dusts to suppress outbreaks or threatened outbreaks of pest arthropods (or weeds). Environmental manipulation (e.g., irrigation practice) is also utilized to enhance activity of certain pathogens and thus effect microbial control, although this is as yet a relatively poorly developed aspect of the tactic.

This chapter deals primarily with microbial control as used against pest insects, and, to a lesser extent, with the control of weeds and plant pathogens. However, it is recognized that the tactic has been used against other types of pests, as has been done with Brazilian *myxomatosis* virus against the European rabbit in Australia (see Chapter 11) and with rust diseases of several weeds.

Indeed, the latter practice is really just in the early phase of what appears to be a very promising exploitation.

History

Insects have long been known to suffer from disease. Aristotle noted that the honey bee, *Apis melifera* L., had diseases, but it was not until 1835 that the silkworm, *Bombyx mori* (L.), was shown to suffer from microbial malaise by Agostino Bassi, who identified a fungus (later named *Beauveria bassiana*) as the causative agent of muscardine disease in that insect. The early studies of insect diseases were largely done on these two "domesticated" insects.

Later, studies of disease were extended to pest insects, and in 1886 a Russian, Krassilstschik, conducted the first field microbial control experiments. He utilized the fungus *Metarrhizium anisopliae* (Metchnikoff) Sorokin against larvae of the sugar beet curculio, *Cleonus punctiventris* Germar, and obtained 50%–80% mortality in his experimental plots. Apparently, the first commercial microbial product available for insect control was produced in Paris in 1891 and contained a fungus [now believed to have been *Beauveria tenella* (Delacroix) Siemaszko]. The insect pathogen most widely used for microbial control today is *Bacillus thuringiensis*, which was first commercially available in France after field experiments begun in the late 1920s showed it was a promising control agent. During the early 1940s a highly successful project involving the control of the Japanese beetle, *Popillia japonica*, by the bacterium *Bacillus popilliae* was initiated by the U.S. Department of Agriculture. This pathogen was later commercially produced and is still in use today. During 1960 in the United States, as a result of the leadership of E. A. Steinhaus, a Food and Drug Administration granted full exemption from tolerance requirements for a commercial preparation of *B. thuringiensis*, which permitted its use on food and forage crops.

The discipline of insect pathology came into full flower immediately after World War II, when E. A. Steinhaus was appointed by H. S. Smith and E. O. Essig to the entomology faculty of the University of California, Berkeley, where he established his famed Laboratory of Insect Pathology. Steinhaus was a dynamic visionary who catalyzed the expansion of insect pathology and its applied branch, microbial control.

But during the 1950s and 1960s microbial control suffered from the excessive enthusiasm of its pioneers as well as overpromotion by the several commercial producers of microbial insecticides. As a result, the tactic lost considerable credibility. In particular, *B. thuringiensis*, whose toxin is the control agent, was poorly exploited and gained an unfortunate reputation as an inferior insecticide. The nuclear polyhedrosis virus of *Heliothis zea* (Boddie) was also badly managed at the research level and failed to live up to initial expectations. But despite these mistakes and setbacks, a group of sound, highly dedicated researchers around the world has maintained its interest in microbial control and is increasingly demonstrating the practical utility of this tactic. Today, *B. thuringiensis* enjoys a solid reputation as an effective and environmentally safe insecticide, and its use is steadily increasing. Virus diseases are also being carefully studied and developed for expanded applied use, as are the entomogenous fungi.

Microbial control is a useful pest management tactic whose importance will surely increase as integrated pest management control programs proliferate, and careful research permits us to adapt the microbial agents to these systems.

Uses and Advantages of Microbial Control Agents

Pathogens have been used in pest control in three basic ways:

1. As introduced agents in classical biological control programs; e.g., Brazilian myxomatosis virus on the European rabbit *Oryctolagus cuniculus*(L.) in Australia (see Chapter 11), nuclear polyhedrosis virus on European pine sawfly, *Neodiprion sertifer* (Geoffroy), in Canada; and *Oryctes* baculovirus (ROV), which has been used successfully against the rhinoceros beetle (*Oryctes* sp.) in the South Pacific.

2. As naturally occurring epizootics incorporated into integrated pest management programs; e.g., use of naturally occurring nuclear polyhedrosis virus on the alfalfa caterpillar (*Colias eurytheme* Boisduval) in California and use of naturally occurring *Entomophthora* spp. (fungi) on the spotted alfalfa aphid, *Therioaphis trifolii*, on alfalfa in California.

3. As sprays or dusts in the nature of microbial insecticides; e.g., *Bacillus thuringiensis* for the control of cabbage looper and many other pest insects and the fungus *Nomuraea rileyi* for velvetbean caterpillar (*Anticarsia gemmatalis* Hubner on soybean.

Microbial agents have a number of advantages as pest control agents—and of course they have disadvantages, too. These advantages and disadvantages are listed in Table 5.1.

The Kinds of Microbial Agents

Viruses

Of the known insect viruses, members of the genus *Baculovirus*, among them the nuclear polyhedrosis (NPV) and granulosis (GV) viruses, have considerable potential for development and use as microbial control agents in agriculture and forestry (Figure 5.1) because of (1) their history of successful

TABLE 5.1. *Advantages and Disadvantages of Microbial Control Agents*

Advantages	Disadvantages
1. Specificity	1. Specificity may be too great when a mixed infestation of closely related pests exists, e.g., *Spodoptera* spp.)
2. Environmental safety	
3. Frequent high virulence to target species	2. Relatively high cost for limited scope.
4. Compatibility with other forms of control (e.g., chemical and biological)	3. Technical and logistical problems in production and application
5. Availability from natural sources (e.g., certain viruses) and their amenability to manipulation as naturally occurring epizootics	4. At present, their nonproprietary nature (i.e., they cannot be patented, as can chemical pesticides)
	5. Sensitivity to physical factors in the environment (e.g., ultraviolet radiation, pH, heat)

use, (2) their high degree of specificity, and (3) their nonhazardous nature relative to nontarget organisms.

Despite their promise and their proven attributes as safe, selective pesticides, there has been only limited use of viruses in arthropod pest control. For example, in the U.S.A., three viral pesticides have been registered with the Environmental Protection Agency and are available for commercial development and sale. These products contain: (1) the NPV of *Heliothis zea*, (2) the NPV of the Douglas fir tussock moth, *Orgyia pseudotsugata* McDonnough, and (3) the NPV of the gypsy moth, *Porthetria dispar* L. The latter two products were registered by the U.S. Forest Service. An NPV originally isolated from the alfalfa looper, *Autographa californica* (Spreyer), is now under study by the U.S. Department of Agriculture, with registration a near possibility. Among the other Western Hemisphere countries, only Canada is making efforts to register viruses as microbial insecticides.

On the other hand, several nonregistered viruses have been used with considerable success by individual growers and government agencies. For example the NPVs of the alfalfa caterpillar and the beet armyworm, *Spodoptera exigua*, have been used by California growers; the NPV of the cabbage looper has been used by growers in California, Canada, and Colombia; the NPVs of several species of sawfly (Tenthredinoidea) have been used by forest entomologists in Canada; the *Oryctes* baculovirus has been used by a governmental agency in the South Pacific; and of course the highly successful myxomatosis virus was used to control the European rabbit in Australia. Experiments conducted by numerous researchers on a variety of pests have shown that a number of additional viruses have considerable promise (see Stairs, 1971, and Pinnock, 1975).

Figure 5.1. A larva of *Hyphantria cunea* infected with nuclear polyhedrosis and granulosis viruses. The gripping of the branch by the hind prolegs is characteristic of virus infections of lepidopterous larvae.

Bacteria

Little is known about the influence of bacteria on insect populations other than through their effects as microbial pesticides applied by man.

At present only two bacteria are currently registered and under commercial production for use in controlling insects. These are *Bacillus popilliae* and *B. thuringiensis* (Figure 5.2). *B. popilliae* has been successfully used against the Japanese beetle in the U.S.A. In this use it is added (mixed) into the soil in large dosages in a grid pattern and then spread through the soil by natural means. This pathogen is expensive to produce, and thus it has been largely used to colonize new areas of infestation. However, employed in this way, it has suppressed outbreaks of the Japanese beetle over wide areas of the U. S. In contrast to *B. popilliae*, *B. thuringiensis* does not occur in nature as an important factor in population regulation. Rather, it is used more as an insecticide, as the crystalline toxin it produces is poisonous to various arthropods (Cooksey, 1971). The strains of *B. thuringiensis* are classified into 12 serotypes that are more or less specific in activity against a spectrum of insects. One strain of *B. thuringiensis* (var. *israelensis*) has an extremely toxic effect on mosquito larvae and yet is harmless to a wide spectrum of nontarget arthropods and other groups in the aquatic environment. Characteristically, the toxins are ingested orally and paralyze the victim. In most cases there are signs of poisoning and a disruption of the midgut epithelium, which allows penetration of bacteria into the haemocoel where septicemia develops (Garcia and Des-Rochers, 1979).

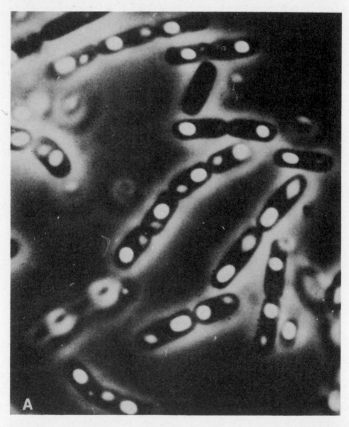

Figure 5.2. Bacteria. (A) Sporangia of *Bacillus thuringiensis*. Note the parasporal crystalline body, which contains an entomotoxin (× 2560) (G. O. Poinar). (B) Sporangia of *Bacillus popillae*, a bacteria used successfully against the Japanese beetle (× 3400) (G. O. Poinar).

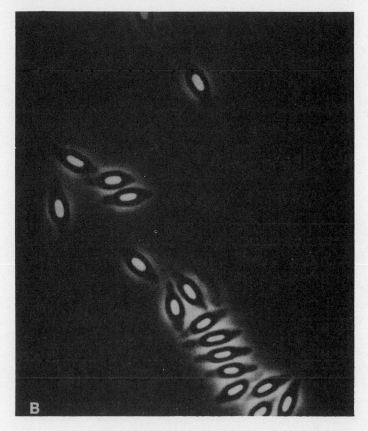

Figure 5.2. *(continued)*

In a recent development, the bacterium *Bacillus sphaericus* Meyer and Neide has also been found to produce toxins that are promising in mosquito control.

Fungi

There are basically two types of fungi infecting insects: (1) a heterogenous group that attacks insects only, and (2) a less specific group that lives on insects as well as on other organic matter.

Completion of the developmental cycle of the entomogenous fungi requires certain conditions of moisture and temperature. This is critical to their performance in nature or under human manipulation. The fungi enter the host mostly from the exterior after contacting and penetrating the cuticle. Transmission of the disease is mainly by conidia, but resting spores, blastospores, and hyphal bodies may also be involved.

Fungal infections in nature may be expressed as massive epizootics, especially if host populations are properly sensitized by abiotic environmental stresses, or the disease may occur only as localized foci. Furthermore, since the fungal spores usually germinate outside their hosts, environmental tolerances are more important than with the other kinds of pathogens that enter through the alimentary canal. Use of fungi in biological (microbial) control must take this into account. Figure 5.3A,B shows two examples of fungal infections.

Success in the deliberate manipulation of fungi has been extremely limited, although there has been much effort. This is essentially because many attempts to use fungi have been conducted by persons who were not trained to recognize the conditions necessary for satisfactory performance of the fungal pathogens. In fact, only a very few pathogenic fungi have been sufficiently well understood from the standpoint of their own characteristics, those of their intended hosts, and those of the conditioning environment to be used with any degree of success (Roberts and Yendol, 1971).

The two most widely used entomogenous fungi are *Beauveria bassiana* and *Metarrhizium anisopliae*. *Hirsutella thompsoni Fisher* has been mass-produced and applied for control of citrus rust mite, *Phyllocoptruta oleivora* (Fishmead). It has provided good control in experimental tests in Florida, Texas, Surinam, and China. In Australia, *Culicinomyces* sp. is now in the safety testing stage prior to large scale use for control of anopheline larvae.

Protozoa

The protozoans appear to have very limited potential as short-term, quick-acting microbial insecticides. For one thing, with few exceptions, they can only be produced on living hosts, which imposes a serious limitation on low-cost, mass-production possibilities.

Second, where injurious or threatening insect infestations occur, few protozoans act with sufficient speed to prevent severe damage to crops. Consequently, if there is to be a place for protozoans in microbial control, it will most likely be where they are introduced into new areas, are spread and maintained by natural means, and reduce the population to a level lower than would occur without them. Cases in point include the successful introduction of the microsporidan *Glugea pyraustae* (Paillot) against the European corn borer, *Ostrinia nubilalis*, in all counties of Illinois by the distribution of infected female moths, and an experiment in France that involved the inoculation of and subsequent spread and increase of *Nosema melolonthae* Krieg in a 12.5-acre plot infested with larvae of the common cockchafer, *Melolontha*

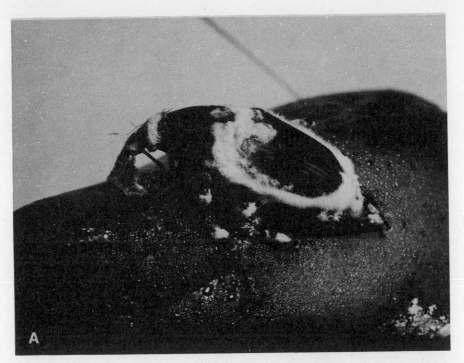

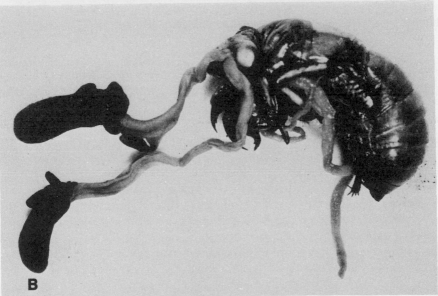

Figure 5.3. Fungi. (A) A banana weevil, *Metamassius hemipterous*, totally infected with *Beauveria bassiana* (G. O. Poinar). (B) Nymph of the cicada, *Diceroprocta apache,* bearing aerial stoma terminated with perithecia of *Cordyceps sobolifera* (G. O. Poinar).

melolontha (L.). *Nosema locustae* Canning has been successfully used for widescale control of grasshoppers in Montana and Idaho rangelands.

However, despite these isolated successes, the use of protozoa remains limited under prevailing microbial control practice.

Nematodea

Nematode parasitization of insects is widespread, and in many instances these round worms are important in controlling insect populations in nature. There is considerable potential for the use of nematodes in the microbial control of insects, but the few attempts made have met with limited success. Difficulties in producing large numbers of these parasites for field inoculation have greatly limited their practical use. If the parasitic species are to be effectively manipulated as microbial control agents, the influence of environmental factors on the nematode dynamics must be better understood, as must nematode behavior and specificity. Currently, then, there is little practical use of parasitic nematodes as microbial control agents against insects, but *Romanomermis culicivorax* Ross and Smith is registered, and the possibility of producing it commercially for control of certain mosquito (anopheline) larvae is being investigated (see Figure 5.4). Species of *Neoaplectana* and *Heterorhabditis* have also been produced in large quantities for field testing against a number of species of Coleoptera and Lepidoptera (Dutky, 1974).

Microbial Control of Weeds and Plant Pathogens

This section derives much of its information from two sources: the current review of the literature on the status of biological control in plant pathology by Schroth and Hancock (1981), and the book *Biological Control of Plant Pathogens* by Baker and Cook (1974). A complete review of this field is best left to experts in plant pathology, and the authorities cited above are highly recommended. The purpose of this section is to illustrate some of the developments in the biological control using microbial pathogens of weeds and pathogens that attack economic plants and to balance the entomological bias of this book.

As Baker and Cook aptly pointed out "Primitive unicellular bacteria and algae are known to have lived in the primeval seas more than three billion years ago," long before plants invaded the land. No doubt, plant pathologists have just begun to scratch the surface of understanding the interrelationships of plants, their pathogens, and the microbial parasites and predators of both that have evolved during these billions of years. Several groups of microorganisms have shown promise in biological control studies, among them fungi,

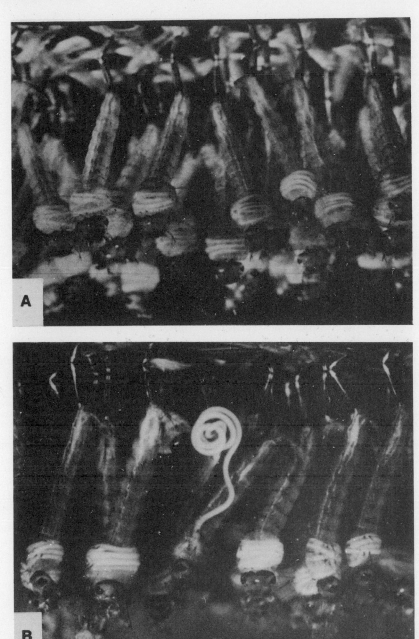

Figure 5.4. Nematodes. (A) Larvae of mosquitoes, *Culex* sp., infected with the mermithid nematode *Romanomermis culicivorax*. (B) An adult nematode exiting from a larva (J. J. Petersen, USDA).

bacteria, viruses, nematodes, phages, and amoebas. Most studies on their effectiveness as biological control agents have been conducted in the laboratory or greenhouse and less often in the field, where there true effectiveness can be tested. Those studies that have been field oriented have been qualitative in nature (Schroth and Hancock, 1981).

Biological Control of Weeds Using Plant Pathogens

Classical biological control of weeds involves the introduction of an exotic pathogen to control an exotic weed recently introduced into a new area. In all respects this procedure has its parallel in classical biological control of insect pests. There have been three outstanding examples using plant pathogens:

1. The control of skeleton weed (*Chondrilla juncea* L.) in Australia by the introduced fungal pathogen *Puccinia chondrillina* Bubak and Syd.
2. The control of wild blackberries (*Rubus constrictus* Mueller and Lefevre and *R. ulmifolius* Schott) in Chile by blackberry rust, *Pharagmidium violaceum* (Schultz) Winter.
3. The control of Pamakani weed (*Agerarina riparia-eupatorium riparia* (nomen nudum)) by the introduced fungal pathogen *Cercosporella agertinae* (nomen nudum).

Often overlooked are the interactions of phytophagous biological control agents and the pathogens they carry passively in the biological control of weeds. A good example of this kind of interaction is that of the moth *Cactoblastis cactorum* (Berg) and several microorganisms in controlling prickly pear (*Opuntia sp.*), which had infested millions of acres of pasture lands in Australia. The interaction of the accidentally introduced elm bark beetle (*Scolytus multistriatus* Marsham) and Dutch elm disease, caused by *Ceratocystis ulmi* (Buisman), in destroying several species of North American elm is an unfortunate but dramatic example of how effective such an interaction can be. As with phytophagous insects, careful screening for host specificity must occur prior to the introduction of pathogens used in biological control—it is work for well-trained specialists.

Augmentation of Native Pathogens to Control Weeds

The application of naturally occurring native pathogens during periods of high host plant vulnerability has been used successfully to kill or suppress weeds. Plant pathologists term pathogens used in this manner *bioherbicides*. Among the successful applications of this technique are:

1. The use of the fungus *Phytophthora citrophthora* (Smith and Smith). Leonian to control milk weed or strangler vine (*Morrenia odorata*), a serious problem in Florida citrus groves because it competes with the tree for sunlight, water, and nutrients and interferes with spraying, harvesting, and irrigation practices. Control of this weed varies from 50% to 77%.

2. The use of the fungus *Colletotrichum gloeosporiodes* f. spp. *Aeschynomene* (Penz) to control northern jointvetch, *Aeschynomene virginica* (L.) B.S.P. in rice fields in the southeastern United States, the degree of control being 90% or better within a month of application.

3. The use of the fungi *Cercospora rodmanii* (Mart.) Solm. and *Uredo eichhorniae* Ciferri and Fragoso for the control of water hyacinth in the United States.

Some of these bioherbicides are or soon will be commercially available in the United States.

Biological Control of Plant Pathogens

In any ecosystem there may be many species, but not all of them interact directly in the food web, and the microflora in soils, waters, and other habitats (e.g., plant surfaces) are no exception. It is relatively easier to observe and study in nature the food web of larger organisms (say, of insects) than it is to study the food web of the microflora. Thus, biological control of plant pathogens is a very difficult endeavor, which must nevertheless be encouraged.

The parasitic habit among members of the microflora is found in phages, bacteria, fungi, and nematodes, which are also often studied by plant pathologists, while predation has been found in amoebae and nematodes. Competition or antibiosis is well documented in the bacteria and fungi. Table 5.2 is a partial list of known field relationships of the biological control of one microorganism by another [summarized from Schroth and Hancock (1981)]; it illustrates the dearth of documented field results in this difficult but very important area of biological control.

The host–natural enemy relationships in the fungi are especially interesting, as the interaction between species may change over time from parasitism to competition, to antibiosis and perhaps mutualism, and may be greatly influenced by the species' physiology and other biotic and abiotic factors. We shall review only two examples of biological control of plant pathogens. Several species of parasitic fungi in the genus *Trichoderma* show good promise as biological control agents of plant pathogens, and some of these are available commercially in Europe, but not in the United States. Among the

TABLE 5.2. *Partial List of Microbial Biological Control Agents of Plant Pathogens*[a]

Biological control agent/method	Host/target organism
Fungi (parasites)	
Trichoderma harzianum Rifai	*Sclerotium rolfsii* Sacc (root diseases) (fungus) *Rhizoctonia solani* Kuehn (damping-off disease) (fungus) *Pythium* spp. (fungus)
Laetisaria sp.	*Pythium ultimum* Trow (seedling diseases) (fungi) *Rhizotonia solani* Kuhn (seedling diseases) (fungi)
Pythium oligandrum Drechsler	*Pythium* spp. (fungus)
Sclerotinia sclerotia *Coniothyrium minitans* Campbell *Sporidesmium sclerotivorum* Vecker, Ayers, and Adams	*Sclerotinia sclerotiorum* (Lib.) Debary (white mold diseases) (fungus)
Bacteria (parasites)	
Bdellovibrio bacteriovorus Stolp and Starr *B. starii* Ceit., Man., and Bap. *B. stolpii* Ceid., Man., and Bap.	Gram-negative bacteria (*Azotobacter chroococcum, Rhizobium* spp.)
Bacterial phages	Wide range of bacteria
Amoebae (predator's)	
Arachnula impatiens Cienkowski	Pathogenic fungal spores in soil, bacteria, blue-green algae, flagellates, and nematodes. Beneficial mycorrhizal fungi.
Manipulated systems	
Agrobacterium radiobacter (strain K-84) (bacterium)	*Agrobacterium tumefaciens* (Sm. and Town, Conn.) (crown gall) (bacteria)
Peniophora gigantea (Fr.) Mass. (fungus)	*Fomes annosus* (Fr.) Cooke (pinestump) (fungus)
Inoculation of lava soil with microflora	*Phytophthora palmivora* Butl. (root rot of papaya) (fungus)
Seed treatments	
Colletotrichum globosum (fungus) *Bacillus subtilis* (Cohn) *Penicillium oxalicum*	Pathogenic soil microflora, exact nature of interaction unknown
Co-evolved systems	
Rhizobacteria, soil bacteria that co-evolved with plants	Displace pathogenic microorganisms on plant roots

[a]Summarized from Schroth and Hancock (1981).

soil pathogens parasitized by *Trichoderma* spp. are Sclerotium rot of peanut and damping-off caused by *Rhizoctonia solani* Kuehn in greenhouses. The recent work by Stirling *et al*. (1978) showing the natural control of root-knot nematode (*Meloidogyne* spp.) in peach orchards in California by the hyperparasitic fungus *Dactylella oviparasitica* Sterling and Mankau is a very important documented example. *D. oviparasitica* attacks nematode eggs and kills the larvae before they hatch. At other times the fungus grows saprophytically on dead root tissues, as well as on the egg masses of other free-living nematodes.

While the biological control of plant pathogens is in its relative infancy, considerable progress has been made using pathogens in classical biological control programs against weeds.

The Future of Microbial Control

Microbial control has considerable potential for expanded application against pest insects, weeds, and plant pathogens. There is also a possibility that the tactic can be successfully applied against plant pathogens (pathogen v. pathogen). It would appear that the greatest potential for expanded use of pathogens lies with the insect viruses. Falcon (1977), in assessing the insect pest problems in Western Hemisphere agriculture, deduced that about 30% of the hemisphere's pest insect species are potentially amenable to treatment with viral pesticides. In California, he estimated that nearly half (46%) of the *major* insect pests are susceptible to baculoviruses alone. His clear implication is that if we make the effort, we can develop important viral insecticides for a significant portion of our pest insects. Collectively, the same probably holds true for the entomogenous fungi, bacteria, and nematodes. It remains to be seen whether the necessary support for this research and development will be found.

References

Baker, K. F., and R. J. Cook. 1974. *Biological Control of Plant Pathogens*. W.H. Freeman: San Francisco. 433 pp.

Cooksey, K. T. 1971. The protein crystal toxin of *Bacillus thuringiensis*: Biochemistry and mode of action. In: H. D. Burges and N. W. Hussey (eds.) *Microbial Control of Insects and Mites*. Academic Press; London. pp. 247–274.

Dutky, S. R. 1974. Nematode parasites. In: F. G. Maxwell and F. A. Harris (eds.) *Proceedings of the Summer Institute of Biological Control of Plant Insects and Diseases*. University Press of Mississippi: Jackson. pp. 576–590.

Falcon, L. A. 1977. Conference on Viral Pesticides: Present Knowledge and Potential Effects on Public and Environmental Health. Myrtle Beach Hilton. Myrtle Beach, Florida. March 21–23, 1977.

Garcia, R. and B. S. DesRochers. 1979. Toxicity of *Bacillus thuringiensis* var. *israelensis* on some California mosquitoes under different conditions. *Mosq. News* **39**:541–544.

Pinnock, D. E. 1975. Pest populations and virus dosage in relation to crop productivity. In: M. Summers, E. R. Engler, L. A. Falcon, and P. V. Vail (eds.) *Baculoviruses for Insect Pest Control: Safety Considerations*. EPA-USDA Working Symposium, Bethesda, Maryland. American Society for Microbiology, Washington, D.C. pp. 145–157.

Roberts, D. W. and W. G. Yendol. 1971. Use of fungi for microbial control of insects. In: H. O.

Burgess and N. W. Hussey (eds.) *Microbial Control of Insects and Mites.* Academic Press : London. pp. 125–149.

Schroth, M. N., and J. G. Hancock. 1981. Selected topics in biological control *Annu. Rev. Phytopathol.* **19** (in press).

Stairs, G. R. 1971. Use of viruses for microbial control of insects. In: H. D. Burgess and N. W. Hussey (eds.) *Microbial Control of Insects and Mites.* Academic Press: London. pp. 97–124.

Stirling, G. R., M. V. McHenry, and R. Mankau. 1978. Biological control of root-knot nematode on peach. *Calif. Agric.* **32:**6–7.

6

Procedures In Natural-Enemy Introduction

Classical biological control was described in an earlier chapter as the control of pest species by introduced natural enemies. The pattern to this process begins with determining whether the target pest is a native or exotic species and then passes through a series of steps involving foreign exploration, quarantine processing of collected material, mass propagation of the natural enemies, their field colonization, and finally the evaluation of their impact on the pest population (see Figure 6.1).

This chapter treats the natural-enemy introduction process in considerable detail, for it is the very essence of classical biological control.

Identification of the Pest as an Exotic Species

In the overwhelming majority of cases, the target pests are exotic species. This circumstance evokes questions: How is a pest recognized as being exotic? How is its native habitat identified? What are the mechanics of exploration once the native habitat has been determined? These questions are addressed and answered in the following pages.

In any contemplated natural-enemy introduction program, it should first be determined whether or not the target pest *is* an exotic invader. There is a dual reason for this: First, if the pest is indeed exotic, the chances for its successful biological control are quite good, but, if it is native, there is considerably less chance that it will be controlled by introducing exotic natural enemies. Second, in determining that the pest is exotic, clues are often found as to its native habitat, which is the logical focus of search for its adapted natural enemies.

A number of indicators serve to identify the exotic invader, the clearest signal being the abrupt outbreak of a species previously unknown in an area.

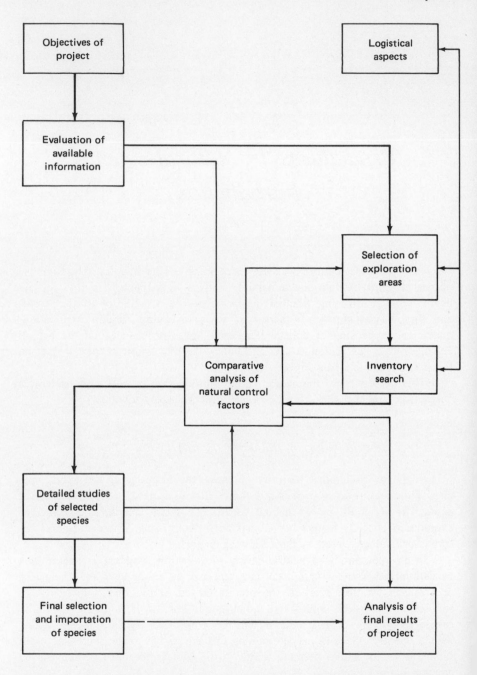

Figure 6.1. Block diagram describing the sequence of procedures and activities followed in a biological control campaign (after Zwolfer *et al.*, 1976).

This is not an infallible criterion, however, since some change in agricultural practice or the introduction of a new crop plant or variety could favor the eruption of a previously obscure indigenous species. But such an eruption is an infrequent occurrence, and a suddenly epidemic native species can usually be recognized, either because it has been previously known in the local fauna, or because it belongs to an endemic taxon.

The really challenging cases of identification involve exotic species that have long resided in an invaded area or one that have become cosmopolitan through accidental transfer by man. For example, most of the pest aphids in the United States are exotic species, but many of them have occurred in our crops for such a long time that they are accepted by farmer and entomologist alike as native species. Furthermore, many are so widely distributed globally that they are considered cosmopolitan and thus presumably Holarctic, when in fact most are really of Palearctic origin.

How then do we recognize the exotic species from among these "old acquaintances" in the pest fauna? For one thing, a pest should be suspect as exotic when it is the sole representative of a taxon whose other members all occur in some remote land. Another strong indication of exotic origin is a pest that lacks important parasites but is heavily attacked in another country; for example, the pea aphid, *Acyrthosiphon pisum*, and the walnut aphid, *Chromaphis juglandicola*, existed for decades in California without important parasites. Heavy parasitization of these aphids by host-specific parasites in the Palaearctic region provided strong evidence that they were native to the Old World. Finally, the host plant affinities of a pest can be an indication that it is an exotic species; for example, in North America the walnut aphid mentioned above is monophagous on *Juglans regia* L. (the so-called English or Persian walnut), an Old World tree, and never occurs on native *Juglans* spp., even though the latter often abounds in the vicinity of commercial *J. regia* groves. This is strong evidence that the aphid originated in the same part of the Old World as did its host tree.

Determination of the Native Habitat of the Exotic Pest

Confirmation that a pest is an exotic species is only one part of the puzzle confronting the biological control specialist. Determination of the pest's area of indigeneity is equally critical, for it is there that its full complement of adapted natural enemies occur. The case of the citrus black scale, *Saissetia oleae*, serves to illustrate this point. This pest was long known to be an invader of California, but for many years its precise native habitat was unknown, since it had a worldwide distribution in tropical and subtropical areas. Thus, an array of ineffective natural enemies was imported into California from widely scattered areas including Brazil and Australia, before the outstanding parasite *Metaphycus helvolus* was obtained from South Africa, the true home of the pest (DeLotto, 1965; van den Bosch, 1968; Annecke and Neser, 1980).

In most cases, the criteria used to identify a pest as an exotic species will reveal its area of indigeneity. Taxonomic and other technical publications, museum material, the pest's host plant affinities, and the opinions of entomologists working with the species will usually, in the aggregate, reveal the true homeland. Our increasing knowledge of insect taxonomy and distribution makes it progressively easier for us to identify the homelands of invading species. But there are still some that are difficult to assign to specific areas of origin. The pink bollworm of cotton, *Pectinophora gossypiella*, is a case in point. With this insect, which occurs in most places on earth where cotton is cultivated, long-accepted opinion holds that India is its native home. But the existence of a complex of *Pectinophora* species in Australia and New Guinea has caused some entomologists to question whether this is true (Ankersmit and Adkisson, 1967). The matter is being investigated, but it is quite apparent that if Australia and New Guinea are indeed the areas to which *P. gossypiella* is indigenous, the current search for parasites may be more promising there than in India.

Some pest species have such vast ranges over contiguous land areas that it is virtually impossible to pinpoint their original habitats. For example, the alfalfa weevil, *Hypera postica*, occurs over an estimated 8–10 million square miles in Europe, Africa, the Middle East, and Asia. This evokes the question: Does all of this vast area represent the true area of indigeneity of *H. postica* or has man over the centuries accidentally spread it from a much more restricted locus as some of its host plants were domesticated and widely planted? Chapter 8 explores this case study in detail

The codling moth, *Laspeyresia pomonella*, the notorious "worm in the apple," is another species whose area of indigeneity in the Palearctic region was originally almost surely more restricted than its distributional range there today. It probably originated in the Himalayan region.

It is very important for us to identify the original centers of distribution of such pests as the alfalfa weevil and codling moth, because it is there that we have the best chance of finding them in association with their full complexes of natural enemies. And with such pests, this becomes something of a guessing game. Thus, we might guess that ancient Media (Northwestern Iran), the original home of alfalfa, may also be the original home of the weevil. But this is a somewhat shaky premise because the weevil attacks other species of *Medicago* and such clovers as *Melilotus* spp., *Trifolium* spp., and *Trigonella* spp., which have different areas of origin. As for the codling moth, we should perhaps look to the original home of the apple, which is presumably somewhere in Central Asia. But here again there is no guarantee that this will be the correct area, since codling moth also attacks other fruits, e.g., apricot, pear, and quince.

Whatever the case, it is of utmost importance that we utilize a variety of criteria in our attempts to determine the native habitat of a given pest (see Zwolfer *et al*, 1976).

Reliance on a single criterion can lead to serious error. For example, when a highly destructive small yellow aphid invaded alfalfa fields in the southwest United States in the middle 1950s, it was identified as *Therioaphis maculata* (Buckton) and designated as the spotted alfalfa aphid. Buckton described *T. maculata* from material collected in India, and if the type locality of the specimens from which he drew his description of this aphid had been the sole criterion upon which the search for its natural enemies was based, the effort would have met with failure—for as it turned out, no parasites of the aphid were found in India. In fact, *Therioaphis maculata* is a synonym of *T. trifolii* (Monell), a species with wide distribution in Europe, Africa, and the Middle East (Hille Ris Lambers and van den Bosch, 1964). It is now quite apparent that this aphid accidentally invaded India and that in doing so escaped its effective parasites, just as it did in invading the southwest United States. Fortunately, multiple criteria (e.g., advice and opinions of entomologists working with *Therioaphis* spp., information on parasites of the group, examination of museum material) were used in establishing the program of exploration for parasites. As a result, collectors were sent to Europe, the Middle East, and Africa, as well as to India. Three species of parasitic Hymenoptera were obtained from Europe and the Middle East, colonized in the southwest United States, and ultimately came to play important roles in the biological control of the aphid (van den Bosch *et al*, 1964).

Importation Agencies in the United States

In the United States seven agencies are intensively involved in exploration for and importation of natural enemies: the Insect Identification and Parasite Introduction Branch of the Science and Education Administration of the U. S. Department of Agriculture (USDA), the University of California, the Hawaii Department of Agriculture, and the Hawaiian Sugar Planters Association. More recently, the University of Florida, Virginia Polytechnic Institute, and Texas A & M University have started to procure natural enemies abroad. Each follows a somewhat different modus operandi, but all have the same objective: the control of insect pests and weeds by imported natural enemies. Basically, each agency places explorers in the field in foreign areas, and these persons then collect promising arthropod parasites, predators and pathogens, and weed-feeding species and transship them to quarantine laboratories where they are tested for desirable biological characteristics before being released to insectaries for mass production and ultimate colonization in the field.

In the United States, all material shipped to the respective quarantine laboratories is by special permit issued by the Plant Protection Division of the USDA. The University of California, the Hawaii Department of Agriculture, the Hawaiian Sugar Planters Association, and the more recently established agencies operate under memoranda of agreement with the USDA that stipulate, among other things, that certain safeguards are to be followed, adequate quarantine facilities provided, and qualified personnel employed in both the collection and quarantine processing of the imported material

Material dispatched from overseas or other distant areas may be transported by several types of carrier, e.g., international airmail, diplomatic pouch, air express. But in all cases, the material is specially packed so as to minimize possible escape of living insects or loss of extraneous plant matter from the parcels.

The efficacy of this "fail-safe" shipping system is evidenced by the fact that although there has been an almost continuous movement of living insect material into the United States by one or another of the biological control agencies since the late 1880s, *there has never been an accident in transit permitting the escape and establishment of an undesirable species.*

As indicated above, the U.S. Department of Agriculture, the University of California, the Hawaii Department of Agriculture, and the Hawaiian Sugar Planters Association each has a somewhat different modus operandi in foreign areas. The U.S. Department of Agriculture personnel operate primarily out of two permanent laboratories in Europe—one near Paris that specializes in the collection and processing of entomophagous insects, and the other in Rome that is concerned with insects and pathogens affecting weeds. At times, U.S. Department of Agriculture personnel are temporarily posted to areas outside Paris or Rome for periods ranging from several weeks up to several years, but most activity is based at the two permanent laboratories.

University of California personnel are permanently stationed in the Divisions of Biological Control at Riverside and Berkeley (Albany) and very recently at Davis, where they are responsible for various projects involving crop pests, forest insects, pests of man and animals, and weeds. Literally all foreign exploration is done by the responsible project leader or project personnel, and all such assignments are of a limited duration (e.g., from several weeks to about one year, except for weed projects which are much longer). These persons usually hold collaboratorships in the U.S. Department of Agriculture and thus (through mutual arrangement) may work out of one of the U.S. Department of Agriculture overseas facilities. At other times, they may make arrangements with foreign institutions (e.g., the Commonwealth Institute of Biological Control, or various universities or ministries of agriculture), or they may carry out their activities completely independently.

In the University of California system, the natural-enemy procurement, propagation, colonization, and evaluation sequence is a totally integrated process under the supervision of the responsible project leader. This is possible because even though California is a large state, its croplands, forests, and other insect-affected areas are readily accessible to project personnel. Thus, once a natural enemy has cleared the quarantine laboratory, its mass propagation, colonization, and eventual evaluation can be easily coordinated by a single person. This gives an essential unity to the whole program.

In the U.S. Department of Agriculture, the entomophagous insects are most often processed through quarantine in Newark, Delaware, and then usually sent to a laboratory located in that part of the country where the pest problem occurs to be mass-propagated there. Even then, material to be released may again be shipped hundreds of miles across state lines to substations or interested researchers if the pest is one that is of regional distribution (e.g., over the corn, wheat, or cotton belts.) Such a scattered operation makes it difficult for a single person to coordinate a program. Thus, once the material has been processed through quarantine, the responsibility for mass propagation, colonization, and evaluation usually rests with one or more of various *other* research branches or persons.

On the other hand, the U.S. Department of Agriculture biological control of weeds program is a highly integrated one. The personnel in the Rome laboratory (and a temporary one in Argentina) collect the candidate insects and conduct studies on their biologies and host plant affinities. Then, when a species has satisfied the basic criteria, it is shipped to the U.S. Department of Agriculture quarantine laboratory at Albany, California, where further testing is conducted. When an insect is cleared for field release, its mass propagation, colonization, and evaluation are carried out by project personnel at Albany or by persons collaborating with these researchers.

The Hawaii Department of Agriculture operation resembles that of the University of California, except that foreign exploration has traditionally been the responsibility of a single individual—a roving, permanent collector—who is on the road full time. Quarantine screening, mass propagation, and colonization are carried out by the Hawaii Department of Agriculture, and the evaluation studies are shared with personnel of the University of Hawaii. The Hawaii Sugar Planters Association operation almost exactly duplicates that of the University of California.

Foreign Exploration for Natural Enemies

The search for and procurement of natural enemies of pest insects is simply specialized insect collecting. The person assigned to a program is invariably a trained entomologist who may or may not have had previous experience in foreign collecting. But, typically, by the time the explorer undertakes a given assignment he will have developed a broad knowledge of the target pest and related species, as well as their known or suspected types of natural enemies. He will be able to recognize the pest in all of its life stages, and he will possess as thorough an understanding of the pest's biology, ecology, phenology, and ethology as possible. In other words, when the explorer moves into the field, he knows what he is looking for and has an excellent idea of where to find it. This is not to imply that the search for natural enemies is an easy task. Indeed it is not. Often the pest is very rare, or it is one of a complex of species and must be distinguished from its congeners. Thus, the discovery and recognition of the pest and its parasites require knowledge, determination, and stamina, particularly if the insects are quite rare or if they occur in remote, inaccessible, or hazardous places.

Depending on the pest species involved and the types of natural enemies being sought, collecting may be done in a variety of environments, e.g., parks, arboretums, streetside vegetation, home gardens, weedy places, orchards, pastures and rangelands, row crops, or undisturbed native vegetation.

The collecting techniques embrace the full spectrum of methods familiar to the trained entomologist: sweeping, beating, rearing (from infested fruits, leaves, stems, or seeds), trapping, handpicking of host insects and/or their parasites themselves.

Desirably, parasite material is shipped in an inactive stage or in a minimally active one (e.g., as pupae, diapausing larvae, developing larvae within parasitized hosts), but sometimes it is necessary to send more active, freely moving forms (e.g., adults). Here moisture and nutrients must often be provided in the shipping container. Predators are also sent in their various developmental stages. Preferably, pupae or other resting stages should be sent, but often it is necessary to send active adults, nymphs, and/or larvae. Here again, nutritional and moisture requirements must be accommodated.

The duration of a foreign exploration program and the area covered will vary with the nature of the problem and with the progress made. Where a search is being undertaken for the first time, it should be programmed to extend over at least one full season's activity of the host species and to cover as much of the pest's distributional range as possible.

With some species this cannot be done in a single season or year, and the natural enemy search may necessarily extend over an indefinite period of time. This is practically true in weed control programs or where a large complex of natural enemies is involved and there appears to be important intraspecific variants of given species. On the other hand, where a particular natural enemy is being sought and its phenology and accepted habitats are understood, the foreign collecting may extend over a period of only a few weeks or even a few days.

Quarantine Reception

The routine introduction of natural enemies from foreign countries for purposes of biological control is fraught with potential hazards. For one thing, it frequently happens—mostly unintentionally, but occasionally by necessity—

that a noxious species, such as a foreign pest, a phytophagous species not hitherto recognized as a pest, or an undetected plant pathogen carried in host plant material or in soil, may arrive in the same shipping parcel with the desired natural enemy. In addition, it is possible for such shipments to contain hyperparasitic species or species that do not attack the phytophagous pests of crops but instead attack the imported primary parasites or predators or natural enemies that already occur in the target region.

The escape of any of these unwanted species could well result in the establishment of a new pest in the target area, a hyperparasite that may interfere with the effective biological control by some indigenous beneficial species, an exotic one already established, or still again, the one to be colonized. Figure 6.2 shows a diagram of a quarantine facility that is designed specifically for processing newly imported natural enemies. The facilities are designed such that light sources are sealed (e.g., by double-pane windows) and positioned so that individuals of phototactic species are attracted away from doors, while access to the facility can be made only through a light-tight vestibule. Insect-proof cages and various types of traps within the facility further ensure that organisms will not be prematurely released.

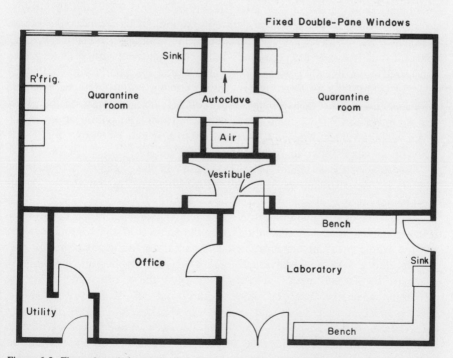

Figure 6.2. Floor plan of Quarantine Laboratory at the Division of Biological Control, University of California, Albany.

It is customary in the United States, for the reception of shipments of beneficial organisms from overseas to be allowed only under permit and then only when assigned to an authorized quarantine laboratory facility. In the United States, these quarantine laboratories are few in number—several operated by the USDA at Moorestown, New Jersey, and Columbia, Missouri (for natural enemies of pest insects); two operated by the USDA in cooperation with the University of California at Albany and Riverside and one in cooperation with the University of Florida (natural enemies of weeds); four operated by the University of California at Albany, Riverside, Davis, and Parlier; one operated by Virginia Polytechnic Institute at Blacksburg (for natural enemies of weeds); one operated by the Hawaii State Department of Agriculture at Honolulu; and more recently, at Texas A & M University.

To protect against accidental introductions of noxious species, a number of activities are carried on in the quarantine laboratory before a desired natural-enemy species is passed out for further propagation and/or colonization. These activities include the separation and rearing by species of adult parasites and predators from the imported material; the careful taxonomic identification of one or more species so recovered; the destruction of any remaining living host specimens, other organisms, and plant material; the culture of recovered beneficial species on locally derived host stocks; and the preliminary working out of the culture requirements and life cycle characteristics of the beneficial species, including observations of its reproductive habits, host selection, sex ratios, physical requirements, and adult food needs. Only when all imported natural-enemy individuals have been identified as to genus, transferred to a local host or prey, and reared in pure culture (thereby ensuring against the obligate hyperparasite) are they allowed to be removed from quarantine. Customarily, such pure cultures are transferred to the insectary for mass, or smaller-scale propagation and eventual colonization. Occasionally, the benefical species is passed on to a research laboratory for further study of its adaptability to local conditions or hosts.

Because of the stringent security required in their operation, quarantine laboratories are restricted as to access. Only authorized quarantine specialists are allowed entry.

The work of the quarantine specialist is of fundamental importance, for it is upon his activities and capabilities that the success of the foreign explorer and the whole program depends. In some cases the foreign explorer is only able to ship a very few individuals of a given entomophagous species with unknown habits and needs. Starting with just these few specimens, the quarantine specialist must be able to carry out all of his examinations, develop handling techniques and culture experiments, and make life history observations. A technical failure involving the loss of a species or a particular strain is costly, not only in time and funds expended, but also from the standpoint of the potential benefit that the introduction might have brought.

Mass Culture of Entomophagous Insects

On occasion, important exotic natural enemies have become established in new areas after only very limited colonizations. For example, the famous *Rodolia cardinalis* became established following the release of but a few hundred individuals. Other examples of this sort are the establishment of *Bathyplectes anurus* (Thompson) on the alfalfa weevil in the eastern United States, and the Iranian strain of *Trioxys pallidus* on the walnut aphid in California.

But despite these "instant" successes, in most cases where species have become established, heavy repeated colonizations were involved. Indeed, the mass production of introduced species is a crucial aspect of any classical biological control program because it permits the release of sufficient numbers of the species at particular colonization sites, provides material for colonization in a variety of environments over a region (i.e., multiple colonizations in space), and allows for repeated colonizations over time.

Another aspect of natural-enemy mass culture involves production of material for inundative (overwhelming) or inoculative (reestablishment) colonizations in sustained programs, such as those practiced in citrus groves in parts of southern California. In these programs, such species as *Cryptolaemus montrouzieri*, a predator of the citrus mealybug, *Metaphycus helvolus*, a parasite of the black scale on citrus and olive, and *Aphytis melinus* DeBach, a parasite of the California red scale on citrus, are maintained on a permanent basis in the insectary and mass cultured as needed for releases in groves where there hosts have risen or threated to rise to injurious levels.

For the several reasons outlined, those organizations that are engaged in biological control maintain insectaries and supporting facilities and staff for the purpose of mass culture of natural enemies (see Figure 6.3). The environ-

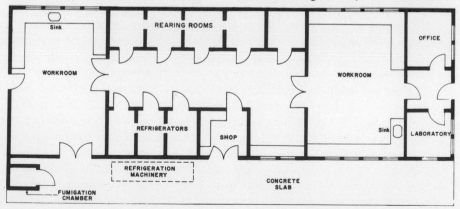

Figure 6.3. Floor plan of a typical mass-culture insectary, modified after that located at the Division of Biological Control, University of California, Albany.

mental conditions in the various rooms of the insectary are varied depending on the requirements of the species involved.

The culture of entomophagous insects and their hosts is a complex matter, and the techniques and manipulations employed necessarily vary with each species being propagated. In the mass propagation process, there are two major concerns: (1) the provision of abundant host material for culturing given natural enemies, and (2) the development of techniques to assure maximum reproductive activity and optimum development and survival of the natural enemies. In other words, techniques must be developed first to propagate the insectary host in sustained volume and then attain efficient propagation of the natural enemy.

Each program poses unique problems that must be worked out. In some cases, easy solutions are reached, while in others, great ingenuity and perseverance are required before satisfactory propagation is attained. Thus, with parasites of certain diaspine scales, long experience has led to the development of a rather standardized rearing technique usually involving use of *factitious hosts* that are amenable to insectary culture. Two scale species that have been particularly important in this role are the latania scale, *Hemiberlesia lataniae* (Signoret), and the oleander scale, *Aspidiotus hederae* (Vallot). In California, these scales, grown on potato tubers or banana squash, have proven to be excellent insectary hosts for parasites of the California red and San Jose scales. This has been of critical importance because these latter scale insects are difficult to mass propagate under insectary conditions and are not biologically optimum hosts in the artificial enviornment. Fortunately, the several parasites of the California red and San Jose scales accept these factitious hosts in the insectary and develop on them. This has permitted efficient mass propagation of these parasites.

Unfortunately, most natural enemies cannot be propagated on other than their natural hosts, which means that techniques must often be developed to culture the pest's host plant, the pest itself, and finally the natural enemy or enemies. Needless to say, such triple-phased programs pose a great challenge to the insectary specialist.

For example, when the decision was made to attempt biological control of the walnut aphid in California, the insectary team was confronted with three major problems: First, they had to develop techniques for maintaining, in good supply, foliated walnut seedlings the year around. Next, they had to develop a program of sustained walnut aphid production. And finally, since at the outset it was unknown just what kind of natural enemies would be imported or how many species, they had to be prepared to quickly develop rearing techniques for whatever was handed them from the quarantine laboratory. The main problem associated with host plant culture was that of maintaining a continuous supply of seedlings in foliage. Walnut is a deciduous tree, and once the plants drop their leaves, they will not regenerate foliage unless conditioned by cold. This was accomplished by holding the dormant seedlings at 38°F for about 6 weeks, after which they were returned to the greenhouse where they then produced new foliage. A continuous supply of foliated seedlings was maintained by rearing the plants in staggered lots (they were grown from nuts planted in vermiculite in pots and watered with a nutrient solution) and rotating these staggered lots from greenhouse, through insectary, through refrigeration, and back to the greenhouse.

The major obstacle to continuous production of aphid and parasite was the potential for each to pass into diapause. This was prevented by maintaining the aphid culture under a 12 hour photoperiod and the parasite culture under constant light. As a result of these manipulations and others of lesser general importance, a highly satisfactory mass propagation program was developed for the parasite *Trioxys pallidus*, several hundred thousands of which were released during the colonization program.

There are, of course, even more formidable problems in insectary propagation than those encountered in the walnut aphid program. For example, at Albany, after several years of effort, we still have not developed sustained production of certain parasites of the walnut husk fly and the alfalfa weevil. Here again, diapause is the key problem. Nevertheless, stocks of these parasites have been maintained and colonizations are being made; they simply aren't being made at high levels. The important point here is that although perseverance, ingenuity, and vigilance have all been applied to the several types of programs described, equal results have not been attained. But the formula works to full effect in most cases, and it is the only way to go about the business of mass propagation of natural enemies.

Colonization

In any biological control program, the entire foreign exploration effort, natural enemy transshipment, and quarantine processing and rearing activities may well go for naught if the candidate species are not properly colonized. Not only is failure at this point frustrating, but it is wasteful of time, energy, and finances. A number of factors militate against successful natural-enemy introduction and these are discussed in Chapter 8. Certain of these factors are beyond human control, and if they are sufficiently adverse they will simply preclude establishment of a given species no matter what degree of care is taken in colonization. But since at its beginning there is no way to know that a program faces impossible odds, any colonization effort must be carried out in such a manner that the natural enemy will have maximum opportunity to become established.

If there is a standard procedure to be applied to natural-enemy colonization it should involve the following: (1) the establishment of optimum conditions for natural-enemy performance at the given colonization site, (2) use of adequate numbers of natural enemies in given releases, (3) a sequence of such releases at each site, and (4) the establishment of colonization sites distributed over the geographical or ecological ranges of the target pest.

With each of these factors there is really no practicable way to determine precisely just what is adequate or optimum. Without question, the most important determinant here is the sustained production of large numbers of natural enemies in the mass propagation insectary. Maximum production of natural enemies will quite obviously permit the colonization of "sufficient" numbers over a period of time at a large number of colonization sites.

In practice, the procedure most commonly followed is to select, in a variety of climatic situations, several typical sites where the pest insect occurs, to note the population dynamics and life history pattern of the pest at these sites, and then to release vigorous stocks of the natural enemy in the immediate vicinity of the appropriate stage or stages of the pest. The number of such sites to be selected and the rates at which the natural enemy is released at each site are determined by the availability of material from the insectary and the rapidity with which recoveries are made.

For example, in an ecological mosaic such as California, where we are currently introducing parasites of alfalfa weevils, there are four distinct climatic areas where intensive colonizations are being made: (1) the intensely hot and arid Colorado Desert Valleys, (2) the mild coastal area, (3) the hot and arid great Central Valley, and (4) the transmontane (Great Basin) area of the state's northeastern corner. In each area, colonization sites have been established in alfalfa on properties of cooperating farmers. Furthermore, in some of the areas, additional sites have been established in uncultivated areas, principally in bur clover, *Medicago hispida* Haertner, a widespread, wild, naturalized host plant of the weevils. Wherever possible, an effort is being made to colonize substantial numbers of the several parasite species in a sequence of releases. The colonizations in alfalfa are being done in collaboration with Agricultural Extension Service personnel who arrange for release sites in commercial alfalfa fields in which the growers allow the colonized areas to remain unmowed and untreated with chemical insecticides through the first two croppings following colonization. Thereafter, normal mowing and control practices are resumed, and the parasites must fend for themselves under these conditions.

The alfalfa weevil parasite colonization strategy will either continue until it has been determined that the parasites have become established or until it is felt that they do not have the ability to do so under the given conditions. In the latter event, more stringently controlled colonization sites may be established, as was done in the program against the spotted alfalfa aphid. In that program, so many hazards faced the released parasites in commercial alfalfa fields that after a year of intensive colonization efforts there was no evidence that any of the three introduced species had established a foothold at any of a large number of colonization sites. Among other things, the repeated mowing of the release fields, frequent chemical treatment of the plots by growers despite prior promises not to do so, and recurrent heavy attacks by coccinellids on the aphid populations (which caused violent fluctuations in aphid numbers) apparently precluded establishment of the parasites (van den Bosch *et al.*, 1959). Consequently, arrangements were made with an alfalfa grower to establish a colonization site in one of his fields, over which complete control could be exercised so that conditions as near optimum as possible for parasite survival and reproduction could be maintained. The grower committed a 38-acre field to the program and gave assurances that under no circumstances would he apply insecticides to any portion of that field. He further agreed to set aside an area of approximately 5 acres as an experimental area in which any type of manipulation or cultural practice deemed necessary to favor establishment of the parasites could be employed. Within the 5-acre experimental area, approximately 1 acre was established as the parasite release plot and therefore left uncut during the entire colonization period. The remaining portion of the 5-acre area was to be used as a parasite "nursery" if and when spread occurred from the 1-acre release point. This was to be accomplished through staggered cutting of the alfalfa, which, it was felt, would favor the continuous occurrence of aphids in the area.

Perhaps the key factor leading to the ultimate establishment of the parasites in this plot was the placing of a large ($6 \times 6 \times 3$ feet) organdy-covered cage in the center of the plot into which the parasites were released. The cage was utilized in order to prevent coccinellid predation on a localized portion of the aphid population. To further favor the parasites, stocks of aphids were periodically introduced into the cage to maintain a consistently high infestation level on the confined alfalfa. Parasites were introduced into the cage at intervals over a 3 month period (April through June, 1955). When abundant mummified aphids were observed on the enclosed plants, the cage was opened to permit spread of the emerging adult parasites from this focus.

In this case, the great care taken in managing the controlled spotted alfalfa aphid colonization site and in manipulating the alfalfa in the surrounding area proved to be the key to the successful establishment of three parasites. By the end of the summer, the parasites had spread into and heavily populated the entire 38-acre field, and they had even moved beyond its boundaries to adjacent fields. In the early autumn, mowed alfalfa bearing millions of parasitized, mummified aphids was trucked to other areas of California and scattered in aphid-infested fields. In this manner, rapid and widespread establishment of the parasites was accomplished from the single source created at the controlled release plot (van den Bosch *et al.*, 1959).

The spotted alfalfa aphid parasite colonization program represents a situation where extreme care and deliberate manipulation were necessary in order to effect establishment of imported natural enemies. The program involving colonization of the Iranian strain of *Trioxys pallidus* against the walnut aphid illustrates the opposite extreme. In this case, very small liberations of *T. pallidus* at a limited number of sites, mostly on streetside and backyard walnut trees, resulted in immediate establishment and rapid increase and spread of the wasp in the season of liberation, even though colonizations were made at a time of year when the aphid population was at its lowest ebb. Furthermore, in the following year, when large-scale colonization of insectary propagated material was planned in commercial walnuts, only a few small, scattered liberations were realized because of an unanticipated breadkown in the insectary program. Nevertheless, the wasp became established everywhere that it was liberated, and it increased explosively (van den Bosch *et al.*, 1970). Within two years of its initial colonization, *T. pallidus* had spread over virtually all of the walnut-growing areas of central and northern California and had attained very important status as a natural enemy of the aphid.

But despite such an easy success, the experience from a majority of the colonization programs suggests that establishment will ordinarily be most difficult to attain, and great care and effort must be expended in the attempt to achieve this goal.

Evaluation of Natural Enemies

The question is often asked, both by the biological control expert and the skeptic, whether any suppression of a host pest population subsequent to the establishment of a biological control agent can be attributed to that natural enemy. In the earlier days of biological control, the simple reduction in pest abundance, that is, the "before and after" pattern of pest density, was commonly cited as the demonstration of natural-enemy efficacy. It was also once customary to use the increase in levels of parasitization or predation by the newly established enemy as a measure of effectiveness. However, it has since been shown that percentage parasitization or predation cannot be equated with level of control exerted because it is the number, not percentage, of survivors that escape natural-enemy action that determines subsequent pest density. The awareness that some new pest invasions and outbreaks eventually subside to lower levels even when no biological control introduction is made—coupled with the criticism by some that biological control rarely worked and that any pest reduction following natural-enemy importation was more likely due to changes in farm practices, in the crop plant, or in the properties of the pest, or to the activities of already established, native natural enemies—led to attempts to evaluate natural-enemy action by experimental and analytical means.

The evaluation of natural-enemy efficacy has followed two patterns. On the one hand, there are experimental procedures, the enemy exclusion or "check" methods that are covered in detail in the following section. The other evaluation approach is the analytical one that makes use of the life table technique. This will be explained in Chapter 7.

Natural-Enemy Exclusion Methods

Four different exclusion methods have been used that can demonstrate the effectiveness of natural-enemy action in particular circumstances. Each has certain limitations or objectionable aspects. As will be seen, the method appropriate to the case must be determined by the nature of the crop, the pest, and the enemies involved (DeBach and Bartlett, 1964; DeBach *et al* 1976).

Mechanical Barrier Method. In this method, a cage, cloth sleeve, wire screen, or similar device is used to enclose an uninfested plant, branch, section of bark, or field plot, and the pest is introduced into the protected zone. An adjacent, infested plot, either not enclosed or enclosed similarly as the above but with openings or other alterations in construction enabling free access of natural enemies, is used as a comparison check. In these situations, the protected pest population almost invariably tends to increase in density, while that in the "unprotected" check plot usually remains at the previous customary low level. The difference in density is ascribed to the action of the excluded natural enemies (DeBach and Bartlett, 1964).

This technique has been applied to black scale on citrus by use of organdy cloth sleeve cages in evaluations of the parasite *Metaphycus helvolus* in southern California; to balsam woolly aphid, *Adelges piceae* Ratz, on fir bark by use of wire screen barriers to evaluate aphid predators; to *Aphis fabae* Scopoli on bean plants by use of plastic screen cages; and to potato aphid, *Macrosiphum euphorbiae* (Thomas), on potato plants to compare the effect of the presence versus absence of coccinellid and syrphid predators on aphid-caused plant damage and tuber yield. The procedure has not been restricted to insects; wire guards have been used to protect barnacles from marine predation in tidal zones.

In certain cases, the barrier exclusion technique has included treatment of the inside surfaces of the barriers to kill off natural enemies trapped inside a cage. The host pest is prevented from coming in contact with the lethal cage, usually by virtue of its sessile (scale or mealybug species) or rather sedentary (aphids, mites) habits.

One objection to the use of cloth or screen cages to protect host infestations from natural-enemy attack is that the internal microclimate of the protected plot may be altered in favor of the host or enemy. So long as barriers tight enough to prevent access of the usually very small natural enemies are used, this objection can only be forestalled by making sure that the "check," or natural enemy plot, has a similar microclimate. This is not always possible. The environment can almost always be expected to be altered such that the efficiency of the natural enemy in controlling the pest is changed.

Chemical Exclusion Method. In this method, a selective pesticide is used to eleminate or greatly inhibit the natural enemy, leaving behind, largely unaffected, the host infestation. Growth of the latter in comparison with a host population in an untreated check plot provides evidence of natural enemy effectiveness (Huffaker and Kennett, 1956; Huffaker *et al.*, 1962; DeBach and Bartlett, 1964).

This technique has been applied successfully to the long-tailed mealybug, the cottony-cushion scale, the California red scale, the yellow scale, and the six-spotted mite on citrus, the cyclamen mite on strawberry, spider mites on apple, olive parlatoria scale on olive, and cabbage root fly on cabbage. It eliminates the objection concerning altered microclimates and cage inter-ference with host movements. However, it has been alleged and even demon-strated in a few cases that the selective pesticide—for example, DDT or its associated spray adjuvants in the case of spidermites, scales, mealybugs and aphids—may very well directly stimulate reproduction and population growth of the host species to the extent that it would diminish or mask any measure of effectiveness of the natural enemy. It has also beeen claimed that the pesticide may stimulate the pest indirectly through a physiological effect on the host plant. This of course can be proven or disproven by appropriate experimentation. A version of this technique, designed to eliminate these objections, makes use of a pesticide barrier or border at the periphery of the study plot. Natural enemies inside the barrier eventually move out to it and are destroyed; enemies outside the barrier zones are prevented from entering the study area. This technique works best with sessile hosts and mobile enemies.

Biological Check Method. This method, applicable only in certain limited situations, is based on the remarkable symbiotic association between certain ants and some of the homopterous pest species. These ants, in exchange for the honeydew excreted by the homopterans (usually certain scale insects, mealybugs, and aphids),"tend" them and protect them from attack by predators and parasites. By eliminating the ants from the infested tree or plant, either with a mechanical barrier or a tanglefoot trap around the base of the trunk or main stem, the pest population is left untended and thereby exposed to full attack by natural enemies. If the natural enemy is an effective one, this results in a decline in the pest density.

There are not questions of altered microenvironment or stimulated host plant responses here, but a question has been raised by the discovery that in certain ant-tended aphid or scale insect species, the ant directly stimulates the reproduction and population growth of the pests. To the extent that this also occurs, ant removal would result in a decline in pest abundance both through increased natural-enemy activity and reduced reproduction rate, making it difficult if not impossible to measure the true effect of the natural enemies.

Hand Removal Method. In certain cases where the pest population is restricted to the plant because of the sessile nature or weak dispersal powers of the species, the removal of all natural enemies from the experimental plant or plant part (e.g., branch) by hand has been attempted. This laborious technique was devised to circumvent the objections that stimulation of the pest population by the manipulation itself rather than the elimination of natural enemies was the cause of the difference between treatment and check. To carry out this method, a group of workers participates on a continuous night and day basis, in the hand removal of any and all natural enemies that settle onto the leaves and twigs of the protected tree or branch. This effort continues for the duration of the test, which must extend long enough to produce a real difference between treatment and check plots.

This technique has been applied to avocado pests in southern California, including lepidopterous caterpillars, mites, mealybugs, and scales (Fleschner, *et al*, 1965) and to cyclamen mite on strawberries (Huffaker and Kennett, 1956).

In many of these enemy-exclusion techniques, the differences in density between the protected and check host populations have been substantial. In some demonstration cages (e.g., California red scale, olive parlatoria/cyclamen mite scale) the increased densities of the protected host populations have led to severe damage or destruction of the crop plant or plant part.

Exclusion Studies in Cotton. Versions of the exclusion methods just described have been used to study the effects of naturally occurring biological control agents in cotton. In one series of experiments, field-grown cotton was placed under large cages, and then the plants in half the cages were treated with a short residue, broad spectrum insecticide to eliminate most of the insects, while those in the remaining cages were not treated and thus remained populated. Next, equal numbers of small bollworm, *Heliothis zea* larvae or eggs were placed on the treated and untreated caged cotton plants, and after a lapse of time their fate was assessed. The analyses showed that where natural enemies (among the spectrum of insect species) were retained, there was strikingly lower survival of bollworms than where the predators had been eliminated (van den Bosch *et al.*, 1969).

In another type of experiment conducted in open fields, several insecticide programs for early and midseason control of *Lygus hesperus* were compared with untreated controls for their effects on natural enemies and subsequent infestations of late-season pests, namely the cabbage looper and the beet armyworm. Here the experiments were large enough (plot size was usually 40 acres and in one case 160 acres) so that interplot movement of natural enemies did not obscure the effects of the various insecticides. The studies clearly showed that where natural enemies were depressed by the *Lygus* control treatments, there were substantial to striking increases in population levels of the lepidopterous pests over those in the untreated controls (Gutierrez *et al.*, 1975; Falcon *et al.*, 1971) (see also Figure 2.4).

References

Ankersmit, G. W. and P. L. Adkisson 1967. Photoperiodic responses of certain geographic strains of *Pectinophora gossypiella* (Lepidoptera). *Insect Physiol.* **13**:553–564.

Annecke, D. P., and S. Neser. 1980. On the black scale *saissetia oleae* (Olivier) and its parasitoids in South Africa. XVI Intntl. Cong. Ent. Kyoto. Japan.

DeBach, P., and B. R. Bartlett .1964. Methods of colonization, recovery, and evaluation. In: P. DeBach (ed.) *Biological Control of Insects and Weeds.* Chapman-Hall: London pp. 412–426.

DeBach, P., C. B. Huffaker, and A. W. MacPhee. 1976. Evaluation of the impact of natural enemies. In: C. B. Huffaker and P. S. Messenger (eds.) *Theory and Practice of Biological Control.* Academic Press: New York, pp. 255–285.

DeLotto, G. 1965. On some Coccidae (Homoptera), chiefly from South Africa. *Bull. Br. Mus. Nat. Hist. (Entomol.)* **16**:175–239.

Falcon, L. A., R. Van den Bosch, J. Gallagher, and A. Davidson. 1971. Investigation of the pest status of *Lygus hesperus* in cotton in central California. *J. Econ. Entomol.* **64**:56–61

Fleschner, C. A., J. C. Hall, and D. W. Ricker. 1965. Natural balance of mite pests in an avocado grove. *Calif. Avocado Soc. Yearb.* **39**:155–162.

Gutierrez, A. P., L. A. Falcon, W. Loew, P. A. Leipzig, and R. van den Bosch. 1975. Analysis of cotton production in California: A model for alcala cotton and the effects of defoliators on its yields. *Environ. Entomol.* **4**:125–136.

Hille Ris Lambers D., and R. van den Bosch. 1964. On the genus *Therioaphis* Walker, 1870, with descriptions of new species (Homoptera: Aphidiae). *Zool. Verh.* No. 68. 47 pp.

Huffaker, C. B., and C. E. Kennett.1956. Experimental studies on predation. I. Predation and cyclamen mite predation on strawberries in California. *Hilgardia* **26**:191–222.

Huffaker, C. B. C. E. Kennett, and G. L. Finney 1962. Biological control of olive scale *Parlatoria oleae* (Clovee) in California by imported *Aphytis maculicornis* (Masi) (Hymenoptera: Aphelinidae). *Hilgardia* **32**:541–636.

van den Bosch, R. 1968. Comments on population dynamics of exotic insects. *Bull. Entomol. Soc. Am.* **14**:112–115.

van den Bosch, R., B. D. Frazer, C. S. Davis, P. S. Messenger, and R. Hom. 1970. *Trioxys pallidus*–An effective new walnut aphid parasite from Iran. *Calif. Agric.* **24**(11)8–10.

van den Bosch, R., T. F. Leigh, D. Gonzalez, and R. E. Stinner. 1969. Cage studies on predators of the bollworm in cotton. *J. Econ. Entomol.* **62**:1486–1489.

van den Bosch R., E. I. Schlinger, J. C. Hall, and B. Puttler. 1964. Studies on succession, distribution and phenology of imported parasites of *Therioaphis trifolii* (Monell) in southern California. *Ecology* **45**:602–621.

van den Bosch, R., E. I. Schlinger, E. J. Dietrick, K. S. Hagen, and I. K. Holloway.1959. The colonization and establishment of imported parasites of the spotted alfalfa aphid in California. *J. Econ. Entomol.* **52**:136–141.

Zwolfer, H., M. A. Ghani, and V. P. Rao. 1976. Foreign exploration and importation of natural enemies. In C. B. Huffaker and P. S. Messenger (eds.) *Theory and Practice of Biological Control.* Academic Press: New York. pp. 189–207.

7

Life Table Analysis in Population Ecology

Given equivalent immigration and emigration, the rates of population births and deaths determine whether populations grow or decline. Immigration and emigration rates are, however, unlikely to be equal, and thus often influence the pattern of population growth. These notions are captured in Figure 7.1. Movements into and out of populations are difficult to assess in nature and are not dealt with directly, but the former would be analogous to births and the latter to deaths in this context.

Various methods of life table analyses have been developed to assess the impact of various sources of intrinsic and extrinsic mortalities on the rate of population growth. Among these methods are age-specific life tables for both laboratory and field, forms of key (mortality) factor analyses, and time-varying life tables. Intuitive discussions of the utility of the different life table methods, their interrelationships, and their applicability to real world problems (e.g., evaluation of biological control programs) are provided in this chapter.

Age-Specific Life Tables

Laboratory Age-Specific Life Tables

Life tables for cohorts or populations are merely schedules of births or deaths caused by various factors. In the laboratory, the intrinsic birth and death rates of a population may be determined under various conditions of food quality, temperature, humidity, and photoperiod. In determining intrinsic parameters, the effects of natural enemies or crowding effects of any kind are usually excluded. A good survey of age-specific life table methods may be found in Andrewartha and Birch (1954). Messenger (1964) and Force and

95

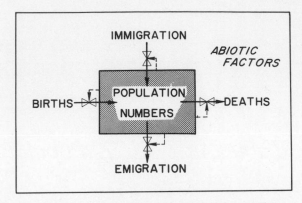

Figure 7.1. Diagram showing the major processes affecting population growth or decline. The direction of the arrows implies direction of change, the symbol ⋈ implies a rate function, and the dotted lines imply density-dependent inhibitions of biological processes.

Messenger (1964) evaluated the performance of the spotted alfalfa aphid, *Therioaphis trifolii* (=*maculata*), and several of its parasites [i.e., *Trioxys utilis* (=*complanatus*) Muesebeck, *Praon exsoletum* (=*palitans*) Nees, Muesebeck, and *Aphelinus semiflavus* (Howard)] under conditions simulating the mild coastal regions and the hot dry Central Valley of California to estimate their ability to survive under these conditions. One of these life tables is given in Table 7.1 along with an explanation of some of the various life table parameters. In general, this type of life table tells more about the physiology of a species (i.e., its survivorship and fecundity patterns over time) than about its ecological performance in the field.

The Appendix lists, for these experimental conditions, the various measures of population success that can be derived from l_x (fraction of original cohort alive at pivotal age x) and m_x (progeny produced per day/ ♀) estimates obtained in the experiments. Among the more interesting parameters of population success are net reproduction rates per female ($R_0 = \Sigma\ l_x m_x$) and the innate capacity for increase (r_m), which can be estimated as

$$r_m \simeq \log_e R_0/\bar{G}$$

where

$$\bar{G} = \Sigma\ x\ l_x m_x/\Sigma\ l_x m_x$$

A more exact method for estimating r_m is given in the Appendix. Figure 7.2A shows the effect of temperature on the average number of progeny per female of the three parasites of the spotted alfalfa aphid, while Figure 7.2B shows the effect on the net reproductive rate. Note that all parasites have the

TABLE 7.1. *Age-Specific Survival and Fecundity Data for P. exsoletum reared at 21°C Mean Temperature, 68% Mean Relative Humidity, and 12-Hour Photoperiod[a]*

Pivotal age (days) (x)	Survival rate (l_x)	Daily fecundity rate: Total eggs (m_x)
0.5	1.00	—
1.5–12.5	1.00	—
13.5	0.93	0.0
14.5	0.93	44.1
15.5	0.93	30.2
16.5	0.93	42.6
17.5	0.93	38.3
18.5	0.93	22.0
19.5	0.93	25.2
20.5	0.89	21.0
21.5	0.89	19.5
22.5	0.89	14.5
23.5	0.89	7.9
24.5	0.81	5.5
25.5	0.72	4.4
26.5	0.64	3.9
27.5	0.51	1.6
28.5	0.42	2.3
29.5	0.34	2.0
30.5	0.30	1.8
31.5	0.26	2.0
32.5	0.04	2.0
33.5	0.04	4.0
34.5	0.04	0.0
35.5	0.04	4.0
36.5	0.00	—

[a]From Messenger (1964).

maximum rate at approximately 22°C. *P. exsoletum* has the lowest fecundity and has a still lower net reproductive rate (R_0) compared to the other two species. R_0 takes into account not only the rate of reproduction but also the survivorship rate, and it amply illustrates the role of mortality in the reproductive potential of a species.

Figure 7.2C shows the effects of temperature on the innate capacity for increase, showing that the aphid potential is greatest at high temperatures—greater than for any of its parasites (*Therioaphis trifolii* > *T. complanatus* > *A. semiflavus* > *P. exsoletum*)—while under cooler conditions the degrees of effectiveness would be different. [But it is important to note that simple comparisons of r_m values between the host and parasite, ignoring the host mortality caused by the parasite, will not tell one whether a parasite might be expected to control the host (see Huffaker, Luck, and Messenger, 1977)].

The simplest population model that can be constructed from such life

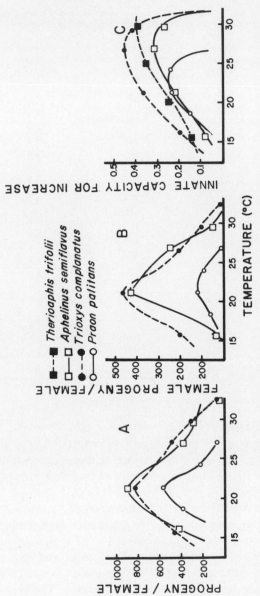

Figure 7.2. Life table parameters of the spotted alfalfa aphid, *Therioaphis trifolii*, and three of its parasites, *Aphelinus semiflavus, Trioxys complanatus*, and *Praon exsoletum* (=*palitans*), as they are affected by temperature. (A) The effects of temperature on progeny per female. (B) The data in A adjusted for female progeny per female. (C) The effects of temperature on innate capacity for increase (r_m). (These figures adapted from Force and Messenger, 1964.)

table data is the well-known Malthusian exponential growth model (i.e., $dN/dt = r_mN$). The number in the population N at time t can be computed from the integral form as

$$N_t = N_0e^{r_mt}$$

where N_0 is the initial population density at time $t = 0$ and r_m is the intrinsic rate of increase. This model assumes unlimited room and food for population growth, and no density-dependent mortality of any kind. Because of the unrealistic nature of these assumptions, ecologists have formulated other simple models (e.g., the logistic model) that assume the environment has some carrying capacity K. The population's intrinsic rate of growth is then modified by its own density:

$$dN/dt = r_mN(1 - N/K)$$

But these models are mainly of theoretical interest and are of limited use in field studies.

In summary, age-specific life tables done in the laboratory give us insight into the physiological response of insects to climatic factors, especially if the experimental conditions used are highly relevant to real climatic regimes (e.g., the bioclimatic studies of Messenger and colleagues), but they often do not tell us whether a natural enemy will control a particular pest. As we shall see, other facets of the species' biology, such as the ability to diapause or aestivate, or mortality caused by biotic agents during some critical period, may be equally important.

Field Age-Specific Life Tables

In ecology we often seek to partition sources of mortality encountered by a species in the field. This analysis requires extensive field sampling of population numbers, keeping in mind to record not only dead ones but also the cause of death. For example, we might find 50 caterpillar larvae, but some may be alive and well and others may be dead. The dead ones may have died because of virus, others from predation, and still others from parasitism. Often we cannot directly see the cause of death, but must infer it from other evidence or perhaps estimate it by use of side experiments. By whatever method we estimate survivorship, we must organize the information in some logical structure, and the most likely form is a life table of the field events. This type of life table has been used widely in studies of north temperate species and is mainly applicable to species with discrete, non-overlapping generations.

An example of such a table, derived from Harcourt (1963), is given in Table 7.2. In this table, the first column specifies the age or stage of development (x) of the pest; the next column the density or number per unit of habitat (N_x) of each of the successive age groups or stages; the third column specifies the various mortality factors (D_xF) acting on each age group; the fourth column lists the numbers killed (D_x) by the corresponding mortality factor; the fifth column gives the percentage dying $(100D_x/N_x)$ during the life stage; while the sixth column gives the percentage dying $(100D_x/N_0)$ based upon the number of individuals (eggs) at the beginning of the generation. At the bottom of the table, departures in the actual sex ratios of emerging adults from the expected male : female ratio of 1 : 1 are entered as adjustments (corrections), either positive or negative, according to whether a lower or higher proportion of males was produced. Losses in the adult stage owing to dispersal, mortality, reduced fecundity deriving from less than average size (the opposite of this enhanced fecundity due to greater than average size would appear as a negative mortality correction) are listed subsequently. The total generation mortality, the expected number of eggs $[E(n + 1)]$ to be produced in the next generation, the counted number of eggs

TABLE 7.2. *Life Table for Diamondback Moth,* [a,b] *Plutella maculipennis (Curtis), on Cabbage, 1951, Ontario, Canada*

Age or stage$_{(x)}$	Density (N_x)	Mortality factor (D_xF)	No. dying (D_x)	Percent of stage dying $(100 \, D_x/N_x)$	Percent of generation dying $(100 \, D_x/N_0)$
Egg	1580	Infertility	25	1.6	1.6
Larva I–IV	1555	1.20-inch rainfall	1199	77.1	77.5
Larva	356	1.52-inch rainfall	36	10.0	79.7
IV	320	*Microplitis plutellae*	52	14.6	83.2
Cocoon (prepupa)	268	*Horogenes insularis*	69	25.7	87.5
Cocoon (pupa)	199	*Diadromus plutellae*	92	46.2	93.3
Adult	107	Sex ratio correction[c]	1	0.9	93.4
Adult	106	Abnormal fecundity[d]	78	73.6	98.3
Adult (normal)	28				

Generation mortality: 99.5%

Expected eggs in next generation: 28/2 × 216 = 3024
Actual eggs in next generation: 864
Trend index: 864/1580 × 100 = 54.7%

[a]Modified from Harcourt (1963).
[b]Sampling unit: Crown quadrant.
[c]Sex ratio correction: Actual SR was 54 males to 53 females. To reduce this to 50 : 50 requires correction of 1% (one male in 107 adults).
[d]Abnormal fecundity: Realized fecundity of a sample of reared females was only 26.4% of normal (216).

$[E(n)]$ at the start of the current generation, and the index of population trend $I = E(n + 1)/E(n)$ complete the table.

The index of population trend indicates whether the population is growing or declining in the next generation. In some life tables, the actual egg density of the following generation is used to compute this value, in which case it specifies the actual population trend rather than the expected. Differences between the expected and actual trend index are considered to reflect immigration (or emigration) of adults into (or out of) the study site.

Thus far, a life table has been explained as a sequence of declining densities of the various stages in a generation, these declines being due to sources of mortality that impinge on the species being studied. The assessment next to be considered is the *manner* in which mortality *changes* from generation to generation or, for species with overlapping generations, from one small time period to the next (e.g., day to day). There are two approaches to these objectives. The first approach requires that we examine a series of age-specific life tables and attempt to deduce the relationship between the intensity of a mortality factor and host density (*mortality factor analysis*); the second approach attempts to decompose the processes of natality–mortality and develop a flexible model of the population dynamics of the species (*models of population dynamics*). Southwood (1978) is suggested reading for a complete discussion of mortality factor analysis, while Gilbert *et al.* (1976) and Gutierrez *et al.* (1979) are suggested for models of population dynamics.

R. F. Morris' Method for Mortality Factor Analysis

Morris (1963a) attempted to estimate species mortality from various sources to final population densities of the spruce budworm, *Choristoneura fumiferana*, in eastern Canada. Indices of population trends $[I = E(n + 1)/E(n)]$ for the egg stage (E) were used to assess whether the populations were growing or declining during generations n to $n + 1$. An approximate form of Morris' model is

$$I = E(n + 1)/E(n) =$$
$$[S_E \cdot S_L \cdot S_P \cdot S_A \cdot P_{\female} \cdot P_f \cdot F \pm [E_m(n)/E(n)]\ ]$$

where S_E, S_L, S_P, and S_A are the within-generation survival rates for the egg, larval, pupal, and adult stages, respectively. The proportion of females ($\female$) in the population is $P_{\female}$, the proportion of realized maximal fecundity (F) is P_f, and E_m is the number of eggs lost or added to the population from the migratory habits of the adults. As we shall see, the various survivorship rates are not constants but depend upon many factors. The above parameters were estimated from field life table studies, in this case, on the spruce budworm.

Morris (1963b) also used regression analyses to determine the relationship between several sources of mortality and indices of population trend estimated from a series of field life table analyses over several consecutive generations. In this manner, he used stepwise multiple regression to predict the effects of various mortality factors on population numbers in one generation [$N(n)$] on the numbers in the next generation [$N(n + 1)$]. His criterion or measure of the extent to which the factor accounted for fluctuations in the population was the increase in coefficient of correlation (i.e., r^2). In theory, a value of $r^2 = 1$ would indicate that all sources of variation had been accounted for, but rarely, if ever, is this possible in the field. Consult the original papers or Southwood (1966) for a more complete review.

There are several disadvantages to this technique: A long period of time is required to collect the data throughout several seasons, the method being applicable only to populations with discrete generations; multiple regression models, while predictive, are not explanatory, and much of the nature of the biological relationships is ignored or, for delayed density-dependent mortality, erroneously assessed; and last, the model is a static description of one data set and tells us little of what would happen if something in the environment (e.g., weather) changed. While we can in hindsight level these criticisms, this work represented a remarkable step forward in insect ecology and furthermore assembled what are probably the best life table data sets for any single species worldwide.

The Varley–Gradwell Method for Mortality Factor Analysis

For species with discrete generations, key factor analysis (Varley and Gradwell,1960, 1963, and 1971) has been used; it is conceptually simpler than the Morris method. To attack a problem using this method, a series of life tables must again be compiled, one for each generation in a sequence of consecutive generations. It is presumed—and experience suggests that the presumption is justified—that in a sequence of generations ranging from 8 to 15 or so the density of the subject species will rise and fall in representative population fluctuations. Such fluctuations will cover a range of densities from high to low, and in this range of densities, the mortality factors acting upon the pest species will exhibit patterns of response to host density. Such patterns, as has been mentioned previously, may be unrelated to density, in which case the mortality factor is classed as *density-independent*; they may be immediately and positively related to density, in which case the mortality factor is classed as *direct density-dependent*; or they may be related to density but only with a time lag, in which case the cause of mortality is classed as *delayed density-dependent*. It is also possible to envision a mortality factor that causes a level of mortality directly but negatively related to host density, and such a factor is called an *inverse density-dependent* factor (see Figure 7.3).

Experience dictates that any of the classes of mortality factors described above can occur at one time or another and in some cases even contemporaneously during a generation of an insect. A sequence of life tables derived from *properly designed sampling schemes* will enable these various patterns to be ascertained. Where introduced natural enemies function as important controlling agents in respect to host density, this life table technique will usually disclose this capability, except at very low host densities.

From the life table data, the numbers (density) of each stage killed are determined and a series of k values are computed (i.e., the log of the numbers killed by the kth factor). The k value for a stage is the difference between the logs of the densities at the beginning and end of the stage. If only one mortality factor was operating, this k value would represent that factor. But if more than one factor was operating, then a separate k is computed for each assignable factor within the life stage, as well as a k value for the residual, unassignable mortality of the stage to complete the stage mortality. Also, a value K, representing total mortality in the generation, is computed as the

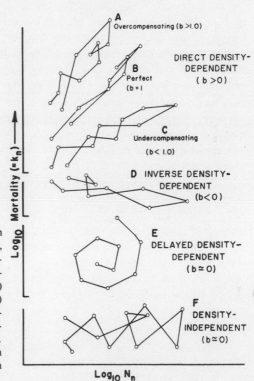

Figure 7.3. The relationships between $\log_{10}$ mortality (k_n) and $\log_{10}$ density, illustrating various types of density-dependent relationships. (A) Overcompensating direct density dependence. (B) Perfect direct density dependence. (C) Undercompensating direct density dependence. (D) Inverse density dependence. (E) Delayed density dependence. (F) Density-independent. (adpated from Southwood, 1966. For further discussion see Varley *et al.*, 1974.)

difference in the logs of numbers of individuals at the beginning of the generation and those surviving to adulthood. Thus, the following equation holds for a species with six stages or age groupings:

$$K = k_1 + k_2 + k_3 + k_4 + k_5 + k_6$$

According to the Varley–Gradwell analysis, if from a series of life tables the different values of k_x are plotted against the log of population density $[\log_{10}(N_x)]$ of age x, the points should define a line of positive slope (Figure 7.3A–C) if the factor causing the k_x mortality is directly density-dependent; a line of zero slope (7.3F) if the factor is density independent; and a line of negative slope (7.3D) if the factor is inverse density independent. A plot in sequence of host generations of k against $\log_{10}N$ is a test of delayed density-dependence of parasite action and spirals counterclockwise and outward in an unstable (Figure 7.4E) interaction (see Southwood, 1966, p. 305).

The level of mortality produced by natural enemies is often not related to the density of the host stage attacked but rather to the host density one generation earlier that produced the number of the searching natural-enemy adults in the current host generation. This delayed density-dependent mortality can be demonstrated by the improvement in the correlation between k and log of host density when k_n is plotted against $\log_{10}(N_x)_{n-1}$ rather than against $\log_{10}(N_x)_n$ where N_n and N_{n-1} are the host densities in the present and the previous generations, respectively. We must also recognize that factors not included in the analysis (e.g., some weather effects) introduce additional variance [see Varley *et al.* (1974) for a review of these methods].

Varley and Gradwell (1960) refer to "key factors" in their analyses of life table data. A key factor is a mortality factor whose effects on the host are the most strongly correlated to changes in total or generation host density and is most easily determined by visually ascertaining which stage k value is most strongly correlated with total K.

Huffaker and Kennett (1969) have warned that life table analyses by themselves, based as they are on correlation rather than demonstrated cause and effect, cannot serve to disclose true natural-enemy effectiveness. Evaluation of effectiveness must be based in part, at least, on experimental evidence. Future evaluations of the importance of natural enemies in bringing about suppression of pest populations will very likely rely on a combination of "check methods" or experimental evidence, and analytical procedures such as life tables and models of population growth.

Time-Varying Life Tables

In general, this section will deal with models that have practical value (i.e., models that are useful in evaluating field populations rather than more theoretical models). However, a short discussion of simple theoretical models is also given here.

Simple Theoretical Models

May (1973) presents a full discussion of applications of theoretical host–parasite models in population dynamics research. These models are not unlike those presented above (Lotka–Volterra models with limiting resources) and are often linked together in predator–prey systems. These models seek to capture the underlying dynamics of predator–prey interactions.

Figure 7.4 shows the relationship of several possible host–parasite sys-

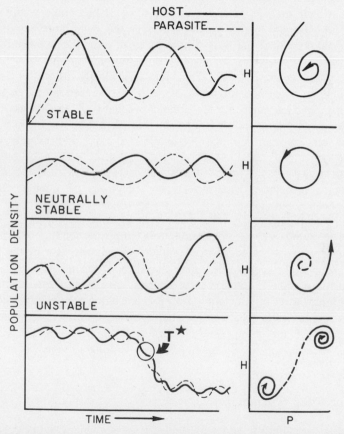

Figure 7.4. Natural-enemy (*P*) and prey (*H*) relationships. (A) A stable *P-H* relationship characterized by oscillations of decreasing amplitude. (B) Neutrally stable relationship characterized by oscillations of equal amplitude. (C) Unstable relationship characterized by oscillations of increasing amplitude. (D) A stable relationship into which a second factor (*T**) has been introduced (e.g., a biocontrol agent) illustrating Holling's concept of resilience. Note that the graphs at the right margin plot corresponding values for *P* and *H*.

tems. In Figure 7.4A, the host–parasite system exhibits damped oscillations, and characteristically, a plot of the values of host (*H*) and parasite (*P*) at various points in time shows a counterclockwise spiral (see the right-hand margin of the figure). Population models showing these characteristics are thought to be stable, i.e., they tend toward an equilibrium level.

The host–parasite populations depicted in Figure 7.4B oscillate with constant amplitude over time and are neutrally stable. By contrast, for those populations shown in Figure 7.4C, the oscillations increase in amplitude, and the population densities are unstable, ending in local extinction of both populations.

The notions of stability are depicted in Figure 7.5A, wherein population numbers can be viewed as a ball in a trough responding to environmental pressures, which include natural enemies. The stable point would be the bottom of the trough, which would represent the equilibrium density. But environmental conditions are rarely constant; most often they change such that the populations may grow, so altering the environment that a new stable region is found (Figure 7.5B). This idea graphically illustrates Holling's (1973) notion of resilience. Figure 7.5C presents the notions of an unstable population interaction such that the populations have no equilibrium density.

These models have not been applied to field situations since all the parameters are constant and they lack age structure and assume that all individuals are equal, but they do give an intuitive feeling of population interactions as well as the notion that populations reach a characteristic level of abundance in nature. The models are however, of dubious value in analyzing field populations.

Turning to Figure 7.4D, we see one host–parasite system interacting at a high level of abundance, as if the parasite is simply responding to changes in its phytophagous host's abundance, the latter being limited not by the parasite but by its own food supply. But after the introduction of an effective, *new* natural enemy at time T^*, the old host population declines and establishes a new equilibrium level in interaction with its controlling new natural enemy. In an applied context this would mean that satisfactory biological control had been achieved if the new pest equilibrium level is below the economic threshold. This notion is an application of Holling's model of resilience (Holling, 1973).

Models of Field Populations

Field age-specific life tables are but snapshots of the dynamics of populations and tell us a little about functional relationships between prey numbers and natural enemy numbers. This is also true of key factor analyses. By contrast, *time-varying life tables* are more like a cinema of the dynamic processes that shape the course of population change and are applicable to populations with either discrete or overlapping generations. Time-varying life tables are mathematical models of the biologies of the natural enemies and

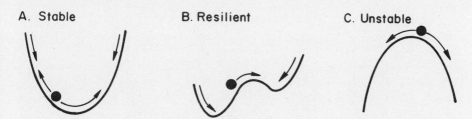

A. Stable   B. Resilient   C. Unstable

Figure 7.5. The notion of a stable point (A) depicted here as the motion of a ball on a wire. Also depicted are the notions of resilience (B) and instability (C).

their prey cast in a dynamic setting. We should note that the model is no better than our knowledge of the biology it describes, hence one should be wary of biology dressed in fancy mathematics—the model should be realistic and should make intuitive sense. Of course, as in a time series of age-specific life tables, the various factors affecting the species' dynamics (e.g., predation, parasitism, weather) change through time. Hughes (1963) first introduced the time-variable life table to ecology in his classic study of the cabbage aphid in Australia. The basic notions are clearly seen in Figure 7.6, which shows the changing proportions over time of the total mortality in the cabbage aphid, *Brevicoryne brassicae* L., attributable to the different mortality factors. Hughes assumed as a necessary condition for his analysis that his aphid populations in the field achieved stable age distributions, an assumption not likely met in species where highly variable general and age-specific mortalities occur. While Carter *et al.* (1978) criticized this obviously weak assumption, Hughes' work was a second remarkable pioneering step toward a realistic method for analysis of insect population dynamics in the field. The method was also able to take into account factors such as changing age structure (i.e., all age classes), different morphs, several trophic levels, time-varying birth and death rates, differences in food quality, weather influences, and the effects of temperature on developmental rates. Because of the complexity, the computations could only be done on a computer (Hughes and Gilbert, 1968).

Conceptually, a time-varying life table can be expressed as a Leslie Matrix (Leslie, 1945), and can be written as

$$
\begin{bmatrix}
N_{1,\,t+\Delta t} \\
N_{2,\,t+\Delta t} \\
N_{3,\,t+\Delta t} \\
\cdot \\
\cdot \\
N_{n,\,t+\Delta t} \\
\cdot \\
N_{m,\,t+\Delta t}
\end{bmatrix}
=
\begin{bmatrix}
b_1 & b_2 & b_3 & \cdot & \cdot & b_n & \cdot & \cdot & & b_m \\
S_1 & \cdot & \cdot & \cdot & \cdot & \cdot & \cdot & \cdot & & 0 \\
0 & S_2 & \cdot & \cdot & \cdot & \cdot & \cdot & \cdot & & 0 \\
0 & & S_3 & \cdot & \cdot & \cdot & \cdot & \cdot & & 0 \\
0 & \cdot & & \cdot & \cdot & \cdot & \cdot & \cdot & & 0 \\
0 & \cdot & & \cdot & \cdot & \cdot & \cdot & \cdot & & 0 \\
0 & \cdot & \cdot & & \cdot & S_n & \cdot & \cdot & & 0 \\
0 & \cdot & & \cdot & \cdot & \cdot & \cdot & \cdot & & 0 \\
0 & 0 & 0 & 0 & 0 & 0 & 0 & S_{m-1} & & 0
\end{bmatrix}_{t}
\times
\begin{bmatrix}
N_{1,\,t} \\
N_{2,\,t} \\
N_{3,\,t} \\
\cdot \\
\cdot \\
N_{n,\,t} \\
\cdot \\
N_{m,\,t}
\end{bmatrix}
$$

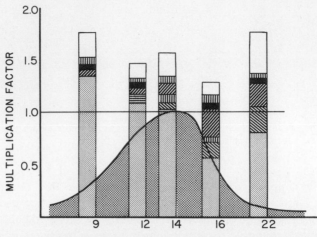

SAMPLING OCCASIONS, PHYSIOLOGICAL TIME SCALE

Figure 7.6. Factors affecting population growth of the cabbage aphid, *Brevicoryne brassicae,* at various times (on a physiological time scale) over the season. The height of the bars depicts the potential multiplication rate of the population, while the lower stippled portions of the bars represent the realized multiplication rate. The other variously shaded parts of the bars represent the realized mortality (e.g., predation, parasitism, disease). The curved line illustrates the rate of population growth (i.e., N_{t+1}/N_t). Physiological time in its simplest form is the day degree concept (adapted from Hughes 1963).

or rewritten as

$$\vec{N}_{t+1} = A_t \cdot \vec{N}_t$$

where N_x are the various cohorts, the elements of A such as b_x are fecundities of the various age (x) classes (called m_x in traditional life table analysis), the S_x are the proportions of each stage that survive to the succeeding stage (equivalent to 1_x), and all other elements are zero. (Note that the components of A_t also vary over time.) In continuous form, these notions are often seen in the literature as

$$\partial N/\partial t + \partial N/\partial a = -\mu\, N(t,a)$$

where a is age, t is time, μ is a complicated birth–death function, and $N(t,0)$ and $N(0,a)$ must be defined; but see Gutierrez *et al.* (1976, 1979) for a discussion of this latter model. We shall only explain the discrete form in detail. The matrix A of age-specific birth (b) and survivorship (S) rates for the m age class are multiplied by the vector of numbers $(\vec{N})$ at time t to estimate the numbers at the next time interval $(t + \Delta t)$. The time interval Δt may be a day, a week, or a year, depending on the species involved. It could also be in some units of temperature-dependent time (degree days).

The computations can be easily seen from the following example of a species with three age classes (note that only age class 3 reproduces):

$$\begin{bmatrix} N_{1,\ t+\Delta t} \\ N_{2,\ t+\Delta t} \\ N_{3,\ t+\Delta t} \end{bmatrix} = \begin{bmatrix} 0 & 0 & b_3 \\ S_1 & 0 & 0 \\ 0 & S_2 & 0 \end{bmatrix} \times \begin{bmatrix} N_{1,\ t} \\ N_{2,\ t} \\ N_{3,\ t} \end{bmatrix}$$

or

$$N_{1,\ t+\Delta t} = 0 \cdot N_{1,t} + 0 \cdot N_{2,t} + b_3 N_{3,t}$$
$$N_{2,\ t+\Delta t} = S_1 \cdot N_{1,t} + 0 \cdot N_{2,t} + 0 \cdot N_{3,t}$$
$$N_{3,\ t+\Delta t} = 0 \cdot N_{1,t} + S_2 \cdot N_{2,t} + 0 \cdot N_{3,t}$$

For the next time step (i.e., $t + 2\Delta t$), we take the values of $\overrightarrow{N}_{t+\Delta t}$ and use them in the right-hand side to estimate $\overrightarrow{N}_{t+2\Delta t}$ on the left-hand side of the equations. Each of the species requires such a model, with the linkages occurring between trophic levels via species (j) specific $b_{x,\ j}$, and $S_{x,\ j}$ values.

In reality, each population could be represented by a model such as shown in Figure 7.7, with the flow of energy between the systems being the death rate of the host and some function of it being the birth rate for the natural enemy (i.e., predator, parasite, or herbivore). We will discuss this energy flow (nutrients) model in a later section.

If the elements in the matrix A are estimated from a laboratory-generated age-specific life table (i.e., they are constant *intrinsic* values), the simulated population would grow exponentially, reaching stable age distributions (i.e., Malthusian growth). This, of course, is highly unrealistic, since population growth causes the environment to deteriorate, hence there is a limit to growth. If, on the other hand, the elements were estimated from field generated age-specific life tables, the fate of the population would depend upon the actual, not intrinsic, birth and death rates. If births are greater than deaths, the population would grow and vice versa. Recall that immigration and migration can be viewed as births and deaths to the population.

But, of course, the birth and death rates also change, and they may be affected by many factors. For example, if births are restricted to the adult stage, the element b_x in A for the immature stages would equal zero. The complexity of the elements of A are illustrated for b_x and S_x below:

$$b_x = b_x(\overrightarrow{N}, F, W, \overrightarrow{Z})$$

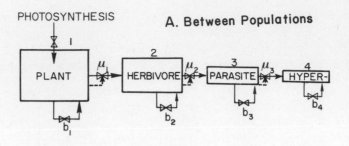

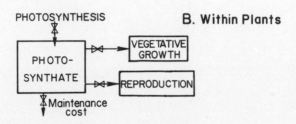

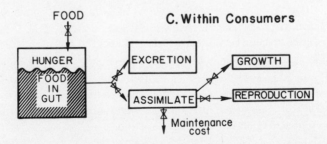

Figure 7.7. Energy flow (A) between populations, (B) within a plant, and (C) within a consumer (e.g., herbivore or predator). The process of assimilation at the individual level illustrates that the processes of respiration, growth, and reproduction are common to all trophic levels.

where $\vec{N}_x$ is the vector of host numbers, F is food quality, W is weather, and $\vec{Z}$ is the series of all other factors.

Similarly, the age-specific survivorship rates might also be complicated functions:

$$S_x = S(\vec{N}, x, P_1, P_2, \ldots P_m, F, W, \vec{Z})$$

where the new variables $P_1, P_2 \ldots P_m$ represent the densities of the m natural enemies. As a detailed essay on estimating b_x and S_x is beyond the scope of this book, interested readers are referred to Gutierrez *et al.* (1976, 1979) and Gilbert *et al.* (1976). Needless to say, both b_x and S_x change radically through time and consequently affect $\vec{N}_x$ or population numbers and the relative proportions in the various age groups. The possible population outcomes are quite varied.

The effects of natural enemies on the survivorship rates are incorporated via predation (or parasitism) submodels such as the "functional response" model developed by Holling (1966) and are part of the S_x in A. The concept of the functional response (Solomon, 1949) was well illustrated by Holling's sandpaper disc experiment. In this famous work, by tapping with the eraser end of a pencil, a blindfolded person searched for small sandpaper discs fastened on to a table. The experiment was carried out for various disc densities and search times. By analogy, the person was the predator and the discs were the prey. The functional response is the relationship of the number of prey killed in an experiment by a single predator as prey density changes (Figure 7.8) and includes such predator attributes as age, consumption, and

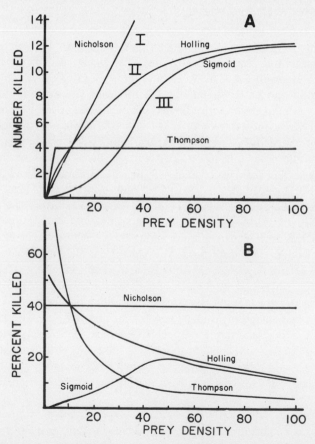

Figure 7.8. Various proposed functional response models depicting (A) the number of prey killed as a function of prey density, and (B) the percentage killed as a function of prey density. (Modified from Huffaker *et al.*, 1968). (See text for further discussion.)

search rates, but can also incorporate such factors as prey size, distributions, and host suitability. The computations are quite laborious.

Theoretically, there are three types of functional responses [Figure 7.8(A)].

1. Type I—The number of prey attacked increases linearly with prey density. This function is unrealistic, as predators have maximum attack rates.
2. Type II—The number of prey attacked increases as a dose response function of prey density. This kind of response is observed for most insect predators and parasites (Holling model).
3. Type III—The number of prey attacked is low at low prey density but because of learning increases with host density until the maximum rate of consumption is reached. This functional response is observed in many vertebrate predators.

A fourth or random search model (Thompson model) assumes that the predator is searching entirely at random. The scale is lower, to enhance clarity. Mathematically, this model would be $1 - \text{PROB} \{0\}$, or the probability of getting one or more attacks. The predicted percent mortality at low densities for this model is very high, but decreases with density (Figure 7.8B). By contrast, the Nicholson model (Type I) predicts a constant percent mortality over all densities. In fact, neither model describes what occurs in nature, where the probability of mortality at low densities has been shown to be intermediate between that predicted by the Thompson and Nicholson models (See Gilbert *et al.*, 1976). The real parameters utilized in these models are not constants in the field, though some were taken as such by the models' authors, but rather they change in value with hunger, prey density and patterns, predator density, and various other factors that influence performance behavior. The only model developed for field use that predicts a reasonable result at low host density as well as a saturation effect at high density is that developed by Frazer and Gilbert (1976) (see Gilbert *et al.*, 1976). It is generally thought that only those functional responses that show an increasing percentage of kill as prey density increases can result per se in prey population regulation (e.g., Type III responses). Several workers have now found Type III functional responses in insect populations in the field because hunger and other physiological factors affect searching behavior and increase capture efficiency as prey host density increases. The percentages of hosts killed are shown in Figure 7.8B.

The predator's (i.e., parasite's or herbivore's) birth rate (i.e., its numerical response) is some function of the prey death rate and is clearly seen as an energy flow model at both the population level (Figure 7.7A) and the individual level (Figure 7.7B,C). Thus, we see that functional and numerical responses are intimately linked and together influence prey population dynamics with immediate and delayed responses, respectively. Note that the general models for the population dynamics of the species and for assimilate allocation are the same for all trophic levels (Gutierrez *et al.*, 1976, 1981).

What Use Are Models?

In practice, these models are not too difficult to develop, but their creation and application require skills in several disciplines, such as biology, ecology, physiology, mathematics, statistics, and of course computer science. These analyses are best conducted by small multidisciplinary research teams. The advantage of good models is that they describe our current state of knowledge about the system in a concise, mathematical form that researchers in other quantitative disciplines can understand. For example, a mathematical model of alfalfa–Egyptian alfalfa weevil interaction (Gutierrez *et al.*, 1976) was used by the economist Uri Regev to evaluate the economics of weevil control, determine optimal decision rules for pest control, evaluate private versus public policy design, and evaluate strategies for the management of insecticide resistance. More importantly, the results of the work are currently being used in pilot projects designed to help farm managers determine the need for pest control as well as the optimal timing for the needed treatments. Farming is a complex business, and various aspects of crop production in addition to pest management pose complex questions in that business. Modeling is a way of organizing the whole system of questions and approaching them in a logical and prioritized way.

On a theoretical level, reasonably complete models can be used to ask questions about the coevolution of interspecies relationships (e.g., Gilbert and Gutierrez, 1973) or to examine subtle details of predator–prey interactions. But we must remember that models are but parodies of nature and can be used to answer questions about nature only to the extent that they capture the relevant natural biology.

Appendix: Life Table Parameters and the Calculation of the Intrinsic Rate of Increase*

The following information is presented as a means of calculating various age-specific life table parameters.

Age-Specific Life Table Parameters

x = Unit of age (e.g., one day)

*See Birch (1948).

q_x = The percentage of the cohort dying in the interval x to $x + 1$ of those alive at age x

l_x = The fraction of individuals of the initial cohort alive at age x

d_x = The fraction of individuals dying during each time interval

L_x = The fraction of individuals alive during the interval between x and $x + 1$ (or the average number number of individuals alive during the interval): $L_x = (l_x + l_{x+1})/2$

T_x = The number of time units lived by the cohort from age x until all individuals die: $T_x = \Sigma L_x$

e_x = The life expectancy of an individual of age x: $e_x = T_x/L_x$

Fertility Life Table Parameters

m_x = The number of offspring produced per ♀ in x

$l_x m_x$ = The number of offspring produced per ♀ in interval x

$\bar{G}$ = Mean generation time

R_0 = The net reproductive rate (or the number of times the population is multiplying each generation under a given set of environmental conditions with unlimited resources). Mathematically this is $R_0 = N_{t + G}/N_t$

We obtain an approximation r_m to the intrinsic rate of natural increase (r_m) in the following manner. Recall that the age specific life table analysis is based on the exponential growth model

$$dN/dt = r_m N$$

The solution to this differential equation is

$$N_t = N_0 e^{r_m(t - 0)}$$

Dividing both sides of this expression by N_0, we see that

$$N_t/N_0 = e^{r_m(t - 0)}$$

If an *average* generation time is $(t - 0) = \bar{G}$, then

$$R_0 = N_{t + \bar{G}}/N_t = e^{r_m \bar{G}}$$

Taking logarithms and dividing

$$r_m = \ln R_0/\bar{G}$$

$\bar{G}$ is estimated by the calculation

$$\bar{G} = \Sigma \ X \ l_x m_x \ / \ \Sigma l_x m_x$$

Another method of computing r_m is that of Laughlin (1965). An alternative form of the exponential growth model is

$$\sum_{x\,=\,0}^{n} e^{-r_m x} \ l_x m_x = 1$$

Using a computer, trial r_m values are substituted into the expression until the left-hand side is (arbitrarily) close to 1.

References

Andrewartha, H. G., and L. C. Birch.1954. *The Distribution and Abundance of Animals.* University of Chicago Press: Chicago.

Birch, L. C. 1948. The intrinsic rate of increase of an insect population *J. Anim. Ecol.* **17**:15–26.

Carter, N., D. P. Aikman, and A. F. G. Dixon. 1978. An appraisal of Hughes' time specific life table analysis for determining aphid reproductive and mortality rates. *J. Anim. Ecol.* **47**:677–688.

Force, D. C., and P. S. Messenger. 1964. Fecundity, reproductive rates, and innate capacity for increase of three parasites of *Therioaphis maculata* (Buckton). *Ecology* **45**:706–715.

Frazer, B. D., and N. Gilbert. 1976. Coccinellids and aphids. *J. Entomol. Soc. B. C.* **73**:33–56.

Gilbert, N. E., and A. P. Gutierrez. 1973. An aphid–parasite plant relationship. *J. Anim. Ecol.* **42**:323–340.

Gilbert, N. E., A. P. Gutierrez, B. C. Fraser, and R. E. Jones. 1976. *Ecological Relationships.* Freeman: London. 157 pp.

Gutierrez, A. P., J. B. Christiansen, C. M. Merrit, W. Loew, C. G. Summers, and W. R. Cothran. 1976. Alfalfa and the Egyptian alfalfa weevil (Coleoptera: Curculionidae). *Can. Entomol.* **108**:635–648.

Gutierrez, A. P., Y. Wang, and R. Daxl. 1979. The interaction of cotton and boll weevil (Coleoptera: Curculionidae)—A study of coadaptation. *Can. Entomol.* **111**:357–366.

Gutierrez, A. P., J. U. Baumgaertner, and K. S. Hagen. 1981. A conceptual model for growth, development, and reproduction in the lady bird beetle, Hippodamia convergens (Coleoptera: Coccinellidae). *Can. Entomol.* **113**:21–33.

Harcourt, D. G. 1963. Major mortality factors in the population dynamic of the diamondback moth *Plutella maculipennis* (Curt.) (Lepidoptera: Plutellidae). *Mem. Entomol. Soc. Can.* **32**:55–66.

Holling, C. S. 1966. The functional response of invertebrate predators to prey density. *Mem. Entomol. Soc. Can.* **48**:3–86.

Holling, C. S. 1973. Resilience and stability of ecological systems. *Annu. Rev. Ecol. Syst.* **4**:1–23.

Huffaker, C. B., R. F. Luck and P. S. Messenger. 1977. The ecological basis of biological control. *Proc. XV Intern. Congr. Entomol.* (Wash. D.C., Aug. 19-27, 1976). pp. 560–586.

Hughes, R. D. 1963. Population dynamics of the cabbage aphid *Brevicoryne brassicae* (L.). *J. Anim. Ecol.* **32**:393–424.

Hughes, R. D., and N. E. Gilbert. 1968. A model of an aphid population—a general statement. *J. Anim. Ecol.* **37**:553–563.

Huffaker, C. B., and C. E. Kennett. 1969. Some aspects of assessing efficiency of natural enemies. *Can. Entomol.* **101**:425–447.

Huffaker, C. B., C. E. Kennett, B. Matsumoto, and E. G. White. 1968. Some parameters in the role of enemies in the natural control of insect abundance. In: T. R. E. Southwood (ed.) *Insect Abundance*. Blackwell: Oxford. pp. 59–75.

Laughlin. R. 1965. Capacity for increase: A useful population statistic. *J. Anim. Ecol.* **34**:77–91.

Leslie, P. H. 1945. On the use of matrices in certain population mathematics. *Biometrika* **33**:183–212.

May, R. M. 1973. On the relationship between various types of population models. *Am. Nat.* **107**:46–77.

Messenger, P. S. 1964. Use of life tables in a bioclimatic study of an experimental aphid–braconid wasp host–parasite system. *Ecology* **45**:119–131.

Morris, R. F. 1963a. The development of a population model for the spruce budworm through the analysis of survival rates. *Mem. Entomol. Soc. Can.* **31**:30–32.

Morris, R. F. 1963b. The analysis of generation survival in relation to age-interval survivals in the unsprayed area. *Mem. Entomol. Soc. Can.* **31**:32–37.

Solomon, M. E. 1949. The natural control of animal populations. *J. Anim. Ecol.* **18**:1–35.

Southwood, T. R. E. 1966. *Ecological Methods with Particular Reference to the Study of Insect Populations*. Methuen: London. 381 pp.

Southwood, T. R. E. 1978. *Ecological Methods with Particular Reference to the Study of Insect Populations*. 2nd ed. Chapman-Hall: London. 500 pp.

Varley, G. C., and G. R. Gradwell. 1960. Key factors in population studies. *J. Anim. Ecol.* **29**:399–401.

Varley, G. C., and G. R. Gradwell. 1963. The interpretation of insect population changes. *Proc. Ceylon. Assoc. Adv. Sci.* **18**:142–156.

Varley, G. C., and G. R. Gradwell. 1971. The use of models and life tables in assessing the role of natural enemies. In: C. B. Huffaker (ed.) *Biological Control*. Plenum Press: New York. pp. 93–112.

Varley, G. C., G. R. Gradwell, and M. P. Hassell. 1973. *Insect Population Ecology: An Analytical Approach*. University of California Press: Berkeley. 212 pp.

8

Factors Limiting Success of Introduced Natural Enemies

Since the initial success against the cottony-cushion scale in California, by 1976 approximately 128 species of pest insects and weeds in many parts of the world have been completely or substantially controlled by imported natural enemies (Laing and Hamai, 1976). Despite this gratifying record, most attempts in classical biological control either have met with total failure, or have been only partially successful (Turnbull and Chant, 1961; Turnbull, 1967; Hall and Ehler, 1979). But this is not reason for despair, as the few limited successes have been of immense value, saving countless millions of dollars for growers and consumers, and have helped reduce pesticide use in agriculture. This record of course needs improvement, and only careful agroeco-system analysis of the factors limiting natural-enemy effectiveness will help show us the way.

In seeking guidelines to improve upon this record, the initial impulse is to look to the highly effective programs and out of them devise some kind of formula that might lead to an improved rate of success in the future. But this is difficult because each program is unique, and the factors that lead to success in one may have little or no bearing on another. Furthermore, it is characteristic of the successful programs that very often the basic reasons for the favorable result are really not known; we simply conclude that we have colonized a highly effective natural enemy or enemies (i.e., they are fully adapted to the target species, vigorous, have excellent searching powers, distribute their progeny efficiently, etc.) under favorable environmental conditions.

But in lieu of a detailed analysis, there are certain general guidelines and practices that if followed or ignored can either enhance or hinder the chances for success in given programs. These guidelines have been learned mostly from programs that failed or fell short of the mark, but in which the reasons for failure were identifiable. Analysis of a number of such programs has indicated that there are certain critical factors that contribute to failure, and it is these factors that we will discuss in this chapter. A qualitative discussion of these several problems is given in this chapter, and a more technical discussion is given in Chapter 12.

Characteristics of the Colonized Environment

The colonized environment is never a duplicate of the habitat from which a natural enemy is obtained, and the degree to which the two differ is of crucial importance to the success or even establishment of the species to be introduced. In other words, in a given program, the physical and biological characteristics of the colonized environment may differ substantially from or closely approach those of the natural enemy's native habitat. This fact bears importantly on the success of the introduced species.

It is generally assumed that the chances for establishment and effective performance of a natural enemy will be greater where the physical environment into which it is being introduced resembles that of the native habitat. For example, one would anticipate that the chances for success of a parasite obtained from the Nile Delta would be greater in the subtropical Imperial Valley of California than in North Dakota. But it should also be emphasized that it is not only the physical nature of the environment that is important but its biotic characteristics as well. Thus, even where the climate of a colonized environment closely resembles that of a natural enemy's native habitat, the enemy may fail because the colonized area lacks a vital alternative host or foodstuff, because the enemy is not adapted to the target pest, or because indigenous hyperparasites may adapt to it and limit its effectiveness.

Certainly the mere presence of the natural enemy's host is no assurance that the enemy will become established or prosper. For one thing, the two have different spectra of requirements, and even though they might share some, the natural enemy may fail because the environment lacks a single factor that is vital to it alone. This situation is reflected in Table 8.1, which lists cases in which characteristics of the colonized environment either precluded the establishment of natural enemies or hindered their fullest performance.

There are, of course, other examples of this sort and other kinds of environmental factors that have precluded the establishment or effective performance of introduced natural enemies, but those listed adequately illustrate the nature of the problem.

Such physical and biological environmental factors are not the only obstacles to success in classical biological control. In fact, other factors such as the kinds of natural enemies colonized, the numbers colonized, and the

Table 8.1. Some Natural-Enemy Species Whose Establishment Was Prevented or Performance Impaired by Adverse Factors in the Colonized Environments

Environmental factor	Country	Natural enemy	Host species	Reference
Adverse climate				
Unfavorable humidity	Mexico	Eretmocerus serius Silv.	Aleurocanthus woglumi Ashby	Clausen (1958)
	U.S. (California)	Aphytis maculicornis (Masi)	Parlatoria oleae (Colvee)	Huffaker et al. (1962)
Temperatures too high	U.S. (California)	Aphytis lignanensis (Masi)	Aonidiella aurantii (Mask.)	DeBach et al. (1955)
	U.S. (California)	Bathyplectes curculionis (Thom.)	Hypera postica (Gyll.)	Michelbacher (1943)
	U.S. (California)	Praon exsoletum Nees, Aphelinus asychis Walk.	Therioaphis trifolii (Monell)	van den Bosch et al. (1964)
	U.S. (central California)	Trioxys pallidus (Halliday) (French strain)	Chromaphis juglandicola (Kalt.)	Messenger and van den Bosch (1971)
Temperatures too low	U.S. (Louisiana)	Several South American parasites	Diatraea saccharalis (Fabr.)	Clausen (1956)
	U.S. (California)	Metaphycus helvolus (Comp.)	Saissetia oleae (Olivier)	Clausen (1956)
	U.S. (Southern California)	Cryptolaemus montrouzieri (Muls.)	Planococcus citri (Risso)	Clausen (1956)
Voltinism	U.S. (interior areas of California where host is even brooded)	Metaphycus helvolus (Comp.)	Saissetia oleae (Olivier)	Clausen (1956)
Lack of alternative host	U.S.	Paradexoides epilachnae Ald.	Epilachna varivestis Muls.	Landis and Howard (1940)
	Australia, France, Italy, Argentina, Brazil and others	Macrocentrus ancylivorus Roh.	Grapholitha molesta (Busck)	Clausen (1958)
Lack of foodstuff	U.S. (eastern)	Tiphia spp.	Popillia japonica Newm.	Clausen (1956)
Lack of competitiveness	U.S. (California)	Aphytis fisheri DeBach	Aonidiella aurantii (Mask.)	DeBach (1965)
Lack of synchrony with host	U.S. (eastern)	Agathis diversus (Mues.)	Grapholitha molesta (Busck)	Clausen (1956)
	U.S. (eastern)	Hyperecteina aldrichi Mesn.	Popillia japonica Newm.	Clausen (1956)

circumstances under which colonizations are made have in the aggregate perhaps been equally detrimental to classical biological control programs.

Poorly Adapted Natural-Enemy Species and Strains

One of the main reasons for the failure of classical biological control programs has been the introduction of natural-enemy species and strains that were poorly or not at all adapted to the hosts against which they were introduced. An illustration of this situation is the great number of natural enemies introduced into California against the black scale, *Saissetia oleae* (van den Bosch, 1968). By 1955 38 species of natural enemies had been introduced against this pest from scattered places around the globe. Of these, 15 became established, but today only one of them, *Metaphycus helvolus*, is of major importance as an enemy of the scale. For years, entomologists collected whatever parasites or predators they found associated with *S. oleae* and allied species, and sent them to California, in the hope that one might be the long-sought effective natural enemy. During these years, parasites were shipped to California from such widely scattered places as Brazil and Tasmania, when in fact the genus *Saissetia* is indigenous to South Africa. It is hardly surprising then, that less than half of the imported black scale natural enemies became established, and that only one of them, a South African species, came to be of any importance.

S. oleae is only one of a number of pest species that have been "shotgunned" with natural enemies in this manner. We have only to look at programs involving such pests as the gypsy moth, the brown-tail moth, the oriental fruit moth, the California red scale, and the European corn borer to find examples of similar broad-spectrum introductions.

In a number of cases, natural enemies of pest insects in one genus have been introduced against species in different genera. This occurred where *Opius* spp. ex the oriental fruit fly, *Dacus dorsalis* Hendel (a tropical species), was colonized on the United States mainland against cherry fruit flies, *Rhagoletis* spp. (a temperate species), and when *Opius* spp. ex Mediterranean fruit fly, *Ceratitis capitata* (an African species), was colonized in California against the walnut husk fly, *Rhagoletis completa* (Cresson) (a Nearctic species). It is hardly surprising that none of these *Opius* became established.

Occasionally an exotic pest's native habitat cannot be determined, and in such cases guesswork is necessarily involved in natural-enemy procurement and introduction. This has occurred with certain of the mealybug and scale insect pests of citrus and has resulted in the considerable colonization of poorly adapted natural-enemy species and races. But guesswork is becoming increasingly rare as our knowledge of the insect faunas and insect systematics and distribution increases, and in the future guesswork should not contribute significantly to the introduction of poorly adapted species. This problem continues to confront biological control workers.

The Importance of Biotypes in Biological Control

A *biotype* is defined here as a subpopulation or race of an organism adapted physiologically and behaviorally to survive under specific climatic conditions of some geographic region. If the biotype is a pathogen, natural enemy, or herbivore, part of the process of natural selection must also include adaptation to the biology of its host's response to climate as well as the host intrinsic physiology. That is to say, interacting species must be adapted not only to local climate but also to each other. These modifications may involve any facet of their biology, such as phenology, host defenses, or site preferences. In a practical biological control sense, any mismatch in the response of an introduced natural enemy and a pest to each other or to climate may cause the introduction to fail or to be only partially successful.

In general, overwhelmingly successful classic biological control projects have involved highly specific or at best narrowly oligophagous parasites and predators. There are several reasons for this. For one thing, many general predators and polyphagous parasites already occur in the invaded habitats and quite obviously do not have a significant or sufficient effect on the introduced species if it continues as an economic pest. So why would imported ones do any better? Second, native pests that already have their adapted natural enemies associated with them are poor targets for exotic natural enemies that most likely have strong biological and behavioral fits by virtue of preadaption of the enemy species to its host or prey. In addition, the situation is further complicated by the presence of climatic biotypes. For example, Campbell *et al.* (1974) showed that the response of the cabbage aphid to temperature varied considerably from Canada to California. This should not be too surprising, as such differences are quite common in nature as is documented by the numerous fruit fly and wheat rust examples. But the high costs and inadequate methodologies for predetermining and assessing the appropriate complex of biological parameters desired in a proposed introduction for it to be most successful are at present unfeasible. Of necessity biological control introductions are a matter of practical urgency, and at present we must admit our ignorance and use intuitive judgments to make decisions as to which species and biotypes to introduce. The role of biotypes in biological control will become self-evident in the following examples.

High percentages of the major pests in many areas are exotic to those areas; e.g.

U.S.A.	+ 50%
California	+ 50%
Australia	Very high
South Africa	Very high

All species are genetically variable, but, as mentioned above, the variability is difficult to assess. In biological control, increasing effort is being made to fit the proper genetic strains or biotypes of natural-enemy species into given biological and ecological situations.

Case Studies

The Alfalfa Weevil

To gain insight into the genetic variation that can characterize a pest species and its natural enemies, we will first refer to the alfalfa weevil, *Hypera postica* (=*H. variabilis*) (Figure 8.1), a Palearctic species that has invaded North America and is now a major pest of alfalfa in the U.S.A., northern Mexico, and Canada. The alfalfa weevil provides a beautiful case in point, for it has an immense distribution in the Palearctic region, over which it occurs under a wide range of ecological conditions. Furthermore, its

Figure 8.1. An adult and a larva of the alfalfa weevil, *Hypera postica* (photo by P. F. Daley).

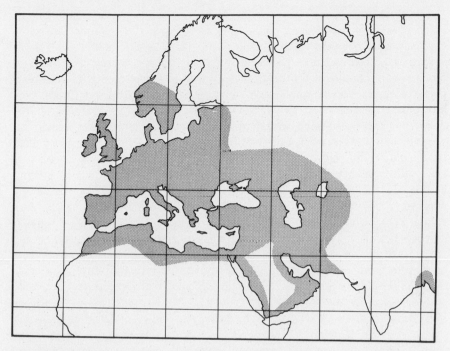

Figure 8.2. Native distribution of the alfalfa weevil, *Hypera postica,* and the Egyptian alfalfa weevil, *H. brunneipennis.*

principal parasite, *Bathyplectes curculionis*, is now virtually co-existent with it.

The alfalfa weevil in its various biotypes (i.e., forms, strains) occurs over about a 12-million-square-mile area extending roughly 4500 miles east to west, and about 2500 miles north to south, ranging west to east from the Atlantic seaboard of Europe to India and Central Russia, and north to south from Sweden to the Sudan (Figure 8.2). This represents not only a tremendous expanse of land but, more importantly, a striking range in climatic and other ecological conditions. The climatic differences between Ahwaz in Khuzistan and Uppsala in Sweden, or between the Cote d'Azur of France and the Kabul Valley of Afghanistan are immense, and yet not only does the alfalfa weevil have the genetic variability to cope with them, but so does its parasite *Bathyplectes curculionis*.

The two species have been able to adapt because they have over the millennia radiated out from their center(s) of origin, probably in the Middle East, and through mutation and selection occupy their current immense and ecologically variable range. Of course, this is not unique to alfalfa weevil; nature is no doubt composed of countless such species.

But when a species is accidentally transported from one continent to another part of that continent, only a very narrow genetically limited introduction may have occurred from a larger pool, like that described above for *H. postica* and *B. curculionis*. In general, the *founder stock* is most likely to be a biotype with a limited and peculiar genetic load in comparison with the whole parent species over its whole area of indigeneity, and different from other biotypes. Unquestionably, many pest biotypes have been accidentally introduced into the wrong kinds of environment and have lacked the capacity to survive. Many other species have no doubt been repeatedly introduced until finally the "right" biotypes were introduced, which then became established pests. Still another species probably made a success of it the first time around. In biological control, biotypes of natural enemies are introduced purposely, but the limitations described above for pests still apply. The natural enemies must be able to establish themselves with regard not only to climate, but also to their host(s) and perhaps other biotic resources as well.

There have been three successful colonizations of U.S. (North America) by *Hypera postica* (Figure 8.3). No one knows how many (if any) unsuccessful colonizations occurred. All evidence indicates that these three successful colonies were derived from distinct biotypes. The first appears to have been from central Europe, and invaded the Salt Lake Valley of Utah early in this

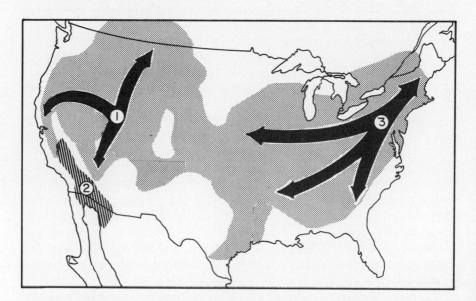

Figure 8.3. Distribution of the alfalfa weevil in the United States. Introductions 1 and 3 were *H. postica,* while 2 was *H. brunneipennis*. Locations of the numbers correspond to original site of introduction.

TABLE 8.2. *Biological and Ecological Differences between the Three Biotypes of Hypera spp. Established in the U.S.A.*

1 (Utah)	2 (Yuma)	3 (Eastern)
All cocooning occurs in ground litter	Some cocooning occurs on aerial parts of plants	Some cocooning occurs on aerial parts of plants
Diapause as scattered individuals in stubble and at field edges	Diapause in aggregations under tree bark, beneath boards, etc.	Diapause as scattered individuals in stubble and at field edges
Cannot exist in hot interior valleys of California	Flourishes in hot interior valleys of California	Not known
Does not encapsulate eggs of *B. curculionis*	Strongly encapsulates *B. curculionis* eggs	Strongly encapsulates *B. curculionis* eggs. In physical appearance resembles Biotype 1 more closely than Biotype 2

century and gradually spread over the Great Basin and then into lower middle California.

The second *H. postica* biotype appears to have come from the Near East (Egypt), entering the United States in the 1930s in the extremely hot Yuma Valley on the Arizona–California border, where the climate is close to that of Khuzistan. The weevil has since spread to other desert valleys of California, the coastal area, and into the great Central Valley of that state, and is everywhere epidemic and a pest of major status. This population is called *H. brunneipennis* by some, but no satisfactory morphological criteria serve to separate it from *H. postica*. There is strong biological evidence (i.e., parasites common to each form) indicating that it is at best a subspecies and is certainly a biotype.

The third *H. postica* invasion occurred on the East Coast of the United States in the 1940s or early 1950s. This population is biologically distinct from the other two populations. However, we do not know its area of origin in the Palearctic Region. The significance of this story is that it reflects the biological and ecological differences that can occur among biotypes of an invading pest species. Table 8.2 summarizes the biological and ecological differences of these biotypes.

What about the parasite *B. curculionis*? There is evidence, too, that the parasite occurs in distinct biotypes. The most clear-cut evidence comes from studies on the *encapsulation* of this species by biotypes of *H. postica* (Salt and van den Bosch, 1966). Encapsulation of the parasite eggs and larvae by host blood cells is an immunity reaction of the host to the parasite. The host recognizes the parasite as foreign rather than as "self". The latter is more common in adapted host–parasite relationships. Intensive ecological studies on all three strains of the weevil have been made, but no obvious differences have turned up, aside from those listed in Table 8.2. For example, their thermal relationships appear to be similar, but obviously strains 1 and 3 must be able to withstand extreme freezing conditions, which would not occur in California.

Because *B. curculionis* is essentially coexistent with *H. postica* in the Palearctic region, it must also have biotypes that are biologically and ecologically adapted to a greater or lesser degree to the whole spectrum of climatic conditions and to Old World weevil biotypes in their various environments.

The Walnut Aphid

In most cases, parasites adapted to different biotypes of pests will not be distinct species but biotypes themselves. A classic case of this sort is that involving the walnut aphid, *Chromaphis juglandicola,* and its parasite, *Trioxys pallidus*. The walnut aphid is native to the western Himalayas, the Hindu Kush, and perhaps parts of Iran—wherever the so-called Persian walnut, *Juglans regia*, is indigenous. The aphid was accidentally introduced in California in the early 1900s and soon became a major walnut pest in California. In the 1950s the parasite *Trioxys pallidus* was collected and imported from France. This species was easily reared in large numbers and released throughout the State, but it failed to become established except in the coastal valleys of Southern California. It apparently could not survive in the major walnut-producing areas in the hot, arid Central Valley (van den Bosch *et al.*, 1979).

Another strain of *T. pallidus* was collected in Iran and shipped to California in 1968. Within two years the Iranian biotype had completely controlled the walnut aphid in all walnut growing areas of California, including the Central Valley. Studies to explain the differences in success between the French and Iranian biotypes of *T. pallidus* were not conducted because of logistic difficulties in handling walnut trees in the laboratory. But a closely related species, *Trioxys complanatus,* and its host, the spotted alfalfa aphid, *Therioaphis trifolii*, are both more manageable in the laboratory, and so it was possible to compare ecological adaptations of the established California *T. complanatus* strains with two biotypes from Rome, Italy, and Karaj, Iran (Flint, 1981).

In these studies, the performance and survival of the three biotypes were assessed at various combinations of temperatures and humidities. The greatest difference was found in the capacity for diapause. The Iranian and Central Valley strains both passed through a diapause under hot or cold extreme temperatures, while the Italian strain lacked this ability. It was not surprising that the Iranian biotype should be better adapted to the Central Valley environment than the Italian biotype when weather regimes of Tehran (Iran) and Rome (Italy) are compared with that of Fresno (California) (Figure 8.4). Similar limitations in the physiology of the walnut aphid parasite possibly explain the success of the Iranian biotype and the failure of the French biotype in the control of the walnut aphid. This case aptly illustrates the importance of the selection of the proper biotype in natural enemy introductions.

Other Examples

Another case involving differential biological adaptation of intraspecific strains of a natural enemy has been reported recently from Canada. This concerns *Mesoleius tenthredinis* Morley, a parasite of the larch sawfly, *Pristiphora erichsonii* Hartig. In this situation a population of the sawfly has reduced the efficacy of the parasite by haemocytic encapsulation of the eggs. A search of Europe revealed the occurrence in Bavaria of an *M. tenthredinis* population that is adapted to the resistant Canadian *P. erichsonii*. The Bavarian *M. tenthredinis* now has become established in Canada and is effecting considerable parasitization in the area of colonization (Turnock and Muldrew, 1971). Canadian entomologists are optimistic that it will eventually spread and take a heavy toll of the sawfly over a wide area.

In addition to the cases cited, there have been other instances where the introduction of better-adapted intraspecific strains of natural enemies has increased the efficacy of biological control programs. Included have been strains of *Aphelinus mali* against woolly apple aphid in China, *Prospaltella perniciosi* Tower against San Jose scale in Switzerland, *Aphytis paramaculicornis* (Masi) against olive parlatoria scale in California, *Comperiella bifasciata* Howard against yellow scale in California, and *Trichopoda pennipes* Fabricius against southern green stink bug in Hawaii.

The several cases just cited clearly point up the importance of the utilization of the proper intraspecific strains of natural enemies in classical biological control programs. Almost certainly, some past failures have re-

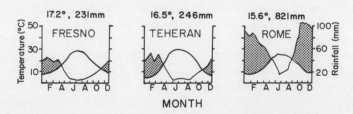

Figure 8.4. Temperature and rainfall regimes of Fresno, California; Tehran, Iran; and Rome, Italy. Shaded areas delimit cool, wet periods of the year. Numbers above the curves represent average annual temperature and total annual rainfall, respectively (from Flint, 1981).

sulted from importation of the wrong strains, and there is a clear need to reassess these failures on the premise that with renewed effort, adapted, effective natural-enemy strains might be found.

But given that control is achieved, it may in fact be fortuitous, since economic control of a pest is man's objective, and not necessarily that of a natural enemy. In a study of the sylvan ecosystem of thimbleberry, *Rubus parviflorus* Nutt., thimbleberry aphid, *Masonaphis maxima* (Mason), and its parasite, *Aphidius rubifolii* Mackauer, Gilbert and Gutierrez (1973) showed that the ecological parameters of the aphid and parasite—and presumably also the plant—were coadjusted such that the parasite took only sufficient toll on the aphid such that both persisted through time. If the parasite (or herbivore) had been more efficient in that ecological context, it would have detrimentally affected its own "fitness," or in other words, lowered its population success. Similar results were found for the interaction of silvan cotton, *Gossypium hirsutum*, and the boll weevil, *Anthonomus grandis* Boh., in central America (Gutierrez and Regev, 1980). Thus nature has a way of "knowing" what it is doing through the long experience of evolutionary trial and error. In dealing with biotypes, we often have to use trial and error methods, but with good field and laboratory research we refine our ideas and increase our rate of successful biological control introductions.

The Mechanics of Natural-Enemy Introductions

When we reflect further upon past biological control efforts it becomes apparent that certain aspects of the mechanics of natural-enemy introductions have doomed many programs to failure. In some cases the persons or agencies involved in the introductions simply did not have the competence to undertake the projects. Then, too, there have been situations where facilities, manpower, or financial support were inadequate to permit the mounting of meaningful efforts. As a result, at times sufficient numbers of natural enemies could not be produced to assure their establishment, and in some cases promising species were lost in culture (e.g., Table 3 in Turnock *et al.* 1976). But even where adequate facilities and skilled staff have been available, some entomophagous insects have proven to be extremely difficult to propagate in the insectary. For example, for years it was impossible to rear certain species of *Coccophagus*, until S. E. Flanders (1937) discovered that the males develop as hyperparasites on females. In other cases, alternate hosts, adult food, symbionts, or some other requisite not found in the new environment may limit the effectiveness of the introduced natural enemy.

Currently, we have another alfalfa weevil parasite, *Bathyplectes anurus*, in our insectary at Albany, California. It has a complex dual diapause; it passes its first diapause as a mature larva in the host cocoon, and while still in the cocoon passes a second diapause as an imago. Thus far, all efforts to prevent induction of diapause in this species have failed, as have efforts to effect early termination of diapause. Thus only one generation of the parasite can be produced per year in the insectary. This of course means that the production of *B. anurus* is a very slow process in the insectary, and as a result its intensive field colonization is hindered. With some natural-enemy species, problems with diapause, mating, and similar biological phenomena have so far defied solution and this has precluded insectary propagation of the species.

Another factor that at times has affected the success of natural-enemy introductions has been indifference on the part of those involved in colonization of the species. This has largely occurred where persons or agencies normally involved in other aspects of insect control have been charged with the responsibility of colonizing parasites or predators shipped to them by one or another of the biological control agencies. It is rather understandable when someone basically involved in or philosophically oriented to some other control tactic does not take a deep interest in the care, colonization, and manipulations of natural enemies shipped to him by some remote laboratory or outside agency. On the other hand, it is inexcusable that we persist in utilizing such a practice in biological control programs.

Conclusion

The preceding examples revealed that a number of factors affect the success or failure of introduced natural enemies. To some it might seem that blind luck as much as any other factor has been the key to the outstanding successes. This, of course, is not true. Koebele went to Australia to search for enemies of cottony-cushion scale because background investigation directed him there. Overwhelming evidence pointed to Australia as the native home of the citrophilus mealybug, and effective natural enemies were indeed found there. A thorough and widespread search produced the Iranian strain of *Aphytis maculicornis* adapted to olive parlatoria scale in California climates. In these cases the right natural enemies were sought in the right places, and they performed with maximum effect in the colonized areas. On the other hand, with the failures or partial successes we are faced with an intriguing question: How many of these cases can be turned into outstanding successes such as that involving the walnut aphid in California and hopefully, too, the larch sawfly in Canada? The odds are that the record can be improved the second time around.

References

Campbell, A., B. D. Fraser, N. Gilbert, A. P. Gutierrez, and M. P. Mackauer. 1974. Temperature requirements of some aphids and their parasites. *J. Appl. Ecol.* 11:431–438.

Clausen, C. P. 1956. Biological control of insect pests in the continental United States. U.S. Dept. Agric. Tech. Bull. 1139. 151 pp.

Clausen, C.P. 1958. Biological control of insect pests. *Annu. Rev. Entomol.* 3:291–310.

DeBach, P. 1965. Some biological and ecological phenomena associated with colonizing entomophagous insects. In H. G. Baker and G. L. Stebbins (eds.) *Genetics of Colonizing
Species.* Academic Press: New York pp. 287–306.

DeBach, P., T. W. Fisher, and J. Landi. 1955. Some effects of meteorological factors on all
stages of *Aphytis lignanensis,* a parasite of the California red scale. *Ecology* 36:743–753.

Flanders, S. E. 1937. Ovipositional instincts and developmental sex differences in the genus
Coccophagous. Univ. Calif. Publ. Entomol. 6(95):401–432.

Flint, M. L. 1981. Climatic ecotypes of *Trioxys complanatus,* a parasite of the spotted alfalfa
aphid. *Environ. Entomol.* 9:501–507.

Gilbert, N.E. and A. P. Gutierrez. 1973. An aphid–parasite plant relationship. *J. Anim. Ecol.*
42:323–340.

Gutierrez, A. P., and U. Regev. 1980. The economic fitness and adaptations in sylvan and
agricultural systems: theoretical and practical applications. Proc. XVI Int. Cong. Entomol.
Kyoto, Japan.

Hall, R.M. and L. E. Ehler, 1979. Rate of establishment of natural enemies in classical
biological control. *Bull. Entomol. Soc. Am.* 25:280–282.

Huffaker, C. B., and C. E. Kennett. 1962. Biological control of the olive scale, *Parlatoria oleae*
(Colvee), in California by imported *Aphytis maculicornis* (Masi) (Hymenoptera: Aphelinidae). *Hilgardia* 32:541–636.

Laing, J. E., and J. Hamai. 1976. Biological control of insect pests and weeds by imported
parasites, predators, and pathogens. In: C. B. Huffaker and P. S. Messenger (eds.) *Theory
and Practice of Biological Control.* Academic Press: New York. pp 685–693.

Landis, B. J., and N. F. Howard. 1940. *Paradexoides epilachnae,* a tachinid parasite of the
Mexican bean beetle. U.S. Dept. Agric. Tech. Bull. 721. 31 pp.

Messenger, P. S., and R. van den Bosch. 1971. The adaptability of introduced biological control
agents. In: C. F. Huffaker (ed.) *Biological Control.* Plenum Press: New York. pp. 68–92.

Michelbacher, A. E. 1943. The present status of the alfalfa-weevil population in lowland middle
California. Calif. Agric. Exp. Sta. Bull. 677. 24pp.

Salt, G., and R. van den Bosch. 1966. The defense reactions of three species of *Hypera*
(Coleoptera, Curculionidae) to an ichneuman wasp. *J. Invertebr. Pathol.* 9:164–177.

Turnbull, A. L. 1967. Population dynamics of exotic insects. *Bull. Entomol. Soc. Am.*
13:333–337.

Turnbull, A. L.,and D. A. Chant. 1961. The practice and theory of biological control of insects
in Canada. *Can. J. Zool.* 39:697–753.

Turnock, W. S., and J. A. Muldew. 1971. *Pristiphora erichsonii* (Htg.), larch sawfly. In:
Bilogical Control Programs against Insects and Weeds in Canada. 1959–1968. *Commonw.
Inst. Biol. Contr. Tech. Commun.* 4:113–127.

Turnock, W. J., K. L. Taylor, D. Schroder, and D. L. Dahlsten. 1976. Biological control of pests
of coniferous forests. In: C. B. Huffaker and P. S. Messenger (eds.) *Theory and Practice of
Biological Control.* Academic Press: New York. p. 289–311.

van den Bosch. R. 1968. Comments on population dynamics of exotic insects. *Bull. Entomol.
Soc. Am.* 14:112–115.

van den Bosch. R., E. I. Schlinger, J. C. Hall, and B. Puttler. 1964. Studies on succession,
distribution and phenology of imported parasites of *Therioaphis trifolii* (Monell) in Southern
California. *Ecology* 45:602–621.

van den Bosch, R., B. D. Frazer, C. S. Davis, P. S. Messenger, and R. Hom 1970. *Trioxys
pallidus.* An effective new walnut aphid parasite from Iran. *Calif. Agric.* 24:8–10.

van den Bosch. R., R. Hom, P. Matteson, B. D. Frazer, P. S. Messenger, and C. S. Davis.
1979. Biological control of the walnut aphid in California: Impact of the parasite *Trioxys
pallidus. Hilgardia* 47:1–13.

9

Analysis of Classical Biological Control Programs

As noted in Chapter 8, 128 or more pest insects and weeds have been completely or substantially controlled by imported natural enemies (Table 9.1). Some of these have been of minor or localized status, while others have been species of continental distribution and great economic importance. In certain cases success was attained with but modest effort, while in others success came only after elaborate preparation, dogged perseverance over many years, and great expense. Among the pest insects, by far the greatest number of successes have been scored against homopterous species, particularly diaspine and lecaniine scales. Some researchers have suggested that scale insects are particularly amenable to biological control because, being sessile during much of their life cycle, they cannot escape or avert natural enemies, and the colonies once found are vulnerable to maximum exploitation. But others point out that scale insects are particularly common pests of horticultural crops such as citrus, and that the considerable success against them may simply reflect the greater emphasis placed on biological control of such orchard pests. There is probably some validity to each contention, but this is perhaps irrelevant, for striking successes have also been scored against species of Coleoptera, Lepidoptera, Diptera, and Hymenoptera in a variety of situations. This is evidence enough that the chances for successful biological control exist across a wide spectrum of the major pest groups.

In the biological control of weeds, there seems to have been disproportionate success against cacti of the genus *Opuntia*. But here again, this simply seems to reflect intensive efforts against this particular pest group. Analysis of the remaining cases listed in Table 9.1 reveals that a wide variety of weedy plant species has been successfully attacked by imported natural enemies. This would indicate that under the right circumstances many additional exotic weed species can be considered as likely prospects for biological control.

TABLE 9.1. Insect Pests and Weeds Substantially or Completely Controlled by Imported Natural Enemies, to 1971[a]

		INSECTS			
Pest species					
Scientific name	Common name	Crop attacked	Place where controlled	Type of natural enemy	Degree of control	Reference

Pest species — Scientific name	Common name	Crop attacked	Place where controlled	Type of natural enemy	Degree of control	Reference
Homoptera						
Acyrthosiphon pisum (Harris)	Pea aphid	Alfalfa	California	Parasite	S[b]	DeBach (1964)
Aleurocanthus spiniferus (Quaint.)	Spiny blackfly	Citrus	Hawaii Japan Guam	Parasite	C[c]	DeBach (1964)
Aleurocanthus woglumi Ashby	Citrus blackfly	Citrus	Cuba Mexico	Parasite	C C	DeBach (1964)
Antonina graminis (Maskell)	Rhodesgrass mealybug	Grasses	Texas	Parasite	S	Schuster *et al.* (1971)
Aonidiella aurantii (Mask.)	Red scale	Citrus	Greece	Parasite	S	DeBach and Argyriou (1967)
Aonidiella citrina (Coq.)	Yellow scale	Citrus	California California	Parasites Parasite	S S	DeBach (1969) DeBach (1964)
Aphis sacchari Zhnt.	Sugarcane aphid	Sugarcane	Hawaii	Parasite and predator	S	DeBach (1964)
Aspidiotus destructor Sign.	Coconut scale	Coconut and other palms	Fiji Mauritius Portuguese W. Africa Bali	Predator Parasite	C	DeBach (1964)
Asterolecanium variolosum (Ratz.)	Golden oak scale	Oak	New Zealand Tasmania	Parasite	S	DeBach (1964)
Cavariella aegopodii (Scop.)	Carrot aphid	Carrot	Australia Tasmania	Parasite	C	van den Bosch (1971)

Species	Common name	Crop	Location	Natural enemy	Result	Reference
... Mask.	... wax scale	Citrus Persimmon Tea	Japan	Parasite	C	DeBach (1964)
Chromaphis juglandicola (Kalt.)	Walnut aphid	Walnut	California	Parasite	S–C	van den Bosch (1971)
Chrysomphalus aonidum (L.)	Florida red scale	Citrus	Israel Seychelles Greece	Parasite Predator Parasite	C S S	DeBach (1964) DeBach (1964) DeBach and Argyriou (1967)
Chrysomphalus dictyospermi (Morgan)	Dictyosperum scale	Citrus	Greece	Parasite	S	DeBach and Argyriou (1967)
Eriococcus coriaceus Mask.	Blue gum scale	Eucalyptus	New Zealand	Predator	C	DeBach (1964)
Eriosoma lanigerum (Hausm.)	Woolly apple aphid	Apple	New Zealand	Parasite	C	DeBach (1964)
Eulecanium coryli (L.)	Vine scale	Grapevine Plum, etc.	British Columbia	Parasite	S	DeBach (1964)
Icerya aegyptiaca Dougl.	Egyptian mealybug or fluted scale	Breadfruit Avocado Citrus	Caroline Islands	Predator	S	DeBach (1964)
Icerya monserratensis Riley and Howard	Fluted scale	Citrus	Ecuador	Predator	S	DeBach (1964)
Icerya purchasi Mask.	Cottony-cushion scale	Citrus	California	Predator and parasite	C	DeBach (1964)
Ischnaspis longirostris (Sign.)	Black thread scale	Fruit and timber trees Coconut palm	Seychelles	Predator	S	DeBach (1964)
Lepidosaphes beckii (Newm.)	Purple scale	Citrus	Mexico Texas	Parasite	S	DeBach (1964)

(continued)

TABLE 9.1. (continued)

INSECTS

Pest species Scientific name	Common name	Crop attacked	Place where controlled	Type of natural enemy	Degree of control	Reference
Nipaecoccus nipae Mask.	Avocado mealybug (Coconut mealybug in Bermuda)	Citrus multiple hosts	Greece Hawaii	Parasite Parasite	S S	DeBach and Argyriou (1967) DeBach (1964)
Orthezia insignis Dougl.	Greenhouse Orthezia	Ornamentals	Kenya	Predator	S	DeBach (1964)
Parlatoria oleae (Colvée)	Olive scale	Olive deciduous fruits	California	Parasite	S–C	DeBach (1964) Huffaker and Kennett (1966)
Perkinsiella saccharicida Kirk.	Sugarcane leafhopper	Sugarcane	Hawaii	Predator	C	DeBach (1964)
Phenacoccus hirsutus Green	Hibiscus mealybug	Multiple	Egypt	Parasite	S	DeBach (1964)
Phenacoccus iceryoides Green	—	Coffee	Celebes	Predator	S	DeBach (1964)
Phenacoccus aceris Sign.	Apple mealybug	Apple	British Columbia	Parasite	C	DeBach (1964)
Pinnaspis buxi (Bouche)	—	Coconut and other palms	Hawaii and Seychelles	Predator	S	DeBach (1964)
Planococcus kenyae	Coffee	Coffee	Kenya	Parasite	C	DeBach (1964)

Host insect	Common name	Crop	Locality	Natural enemy	Result	Reference
...pentagona Targ.	scale	Mulberry, Papaya, Oleander, etc.	Puerto Rico	Parasite	s	DeBach (1964)
			Bermuda	Predator	S	DeBach (1964)
Pseudococcus spp.	Mealybug	Mulberry	Georgian S.S.R. (USSR)	Parasite	S	Kobakhidze (1965)
Pseudococcus citriculus Green	—	Citrus	Australia	Predator	S	DeBach (1964)
Pseudococcus comstocki (Kuw.)	Comstock mealybug	Apple	Eastern Russia	Parasite	C	DeBach (1964)
			California	Parasite	C	Kobakhidze (1965)
Pseudococcus gahani Green	Citrophilus mealybug	Citrus	California	Parasite	C	DeBach (1964)
Quadraspidiotus perniciosus Comst.	San Jose scale	Apple and other fruits	Chile	Parasite	S	DeBach (1964)
			Switzerland	Parasite	S–C locally	Mathys and Guignard (1965)
Saissetia oleae (Olivier)	Black scale	Citrus; Olive	California	Parasite	S	DeBach (1964)
Saissetia nigra (Nietn.)	Nigra scale	Olive	Peru	Parasite	S	DeBach (1964)
Siphanta acua (Wlk.)	Torpedo bug; Planthopper	Ornamentals; Several hosts	California; Hawaii	Parasite	S	DeBach (1964)
Tarophagus proserpina (Kirk.)	Taro leafhopper	Taro	Hawaii	Predator	S	DeBach (1964)
Therioaphis trifolii (Monell)	Spotted alfalfa aphid	Alfalfa	California	Parasite	S	DeBach (1964)
Trialeurodes vaporariorum (Westw.)	Greenhouse whitefly	Tomatoes, etc.	Australia	Parasite	S	DeBach (1964)
Trionymus saccharis (Ckll.)	Pink sugarcane mealybug	Sugarcane	Tasmania	Parasite	S	DeBach (1964)
			Hawaii	Parasite	S	DeBach (1964)

(continued)

TABLE 9.1. (continued)

INSECTS

Pest species — Scientific name	Common name	Crop attacked	Place where controlled	Type of natural enemy	Degree of control	Reference
Lepidoptera						
Cnidocampa flavescens (Wlk.)	Oriental moth	Shade trees	Massachusetts	Parasite	S	DeBach (1964)
Coleophora laricella (Hbn.)	Larch casebearer	Larch	Canada	Parasite	S	DeBach (1964)
Diatraea saccharalis (F.)	Sugarcane borer	Sugarcane	West Indies	Parasite	S	DeBach (1964)
Grapholitha molesta (Busck.)	Oriental fruit moth	Peach	Canada	Parasite	S	DeBach (1964)
Harrisina brillians B&McD.	Western grapeleaf skeletonizer	Grape	California	Parasite	S	DeBach (1964)
Homona coffearia Nietn.	Tea tortrix	Tea	Ceylon	Parasite	C	DeBach (1964)
Laspeyresia nigricana (Steph.)	Pea moth	Vegetables	British Columbia	Parasite	S	DeBach (1964)
Levuana iridescens B.B.	Coconut moth	Coconut	Fiji	Parasite	C	DeBach (1964)
Nygmia phaeorrhoea (Donov.)	Brown-tail moth	Deciduous forest, and shade trees	Northeast U.S.	Parasite	S	DeBach (1964)
Operophthera brumata (L.)	Winter moth		Canada Nova Scotia	Parasite	C C	DeBach (1964) Embree (1971)
Pieris rapae (L.)	Imported cabbage worm	Cruciferous crops	New Zealand	Parasite	S	DeBach (1964)

	common name	crops / Forest trees	location		status	reference
Stilpnotia salicis (L.)	Satin moth	Forest trees	U.S. British Columbia	Parasite	S	DeBach (1964)
Coleoptera						
Anomala orientalis Waterh.	Oriental beetle	Sugarcane	Hawaii	Parasite	S	DeBach (1964)
Anomala sulcatula Burm.	—	Sugarcane	Saipan	Parasite	C	DeBach (1964)
Brontispa longissima selebensis Gestro	Coconut leaf miner	Coconut	Celebes	Parasite	S	DeBach (1964)
Brontispa mariana Spaeth	Mariana coconut beetle	Coconut	Mariana Islands	Parasite	S	DeBach (1964)
Gonypterus scutellatus Gyll.	Eucalyptus weevil	Eucalyptus	South Africa	Parasite	C	DeBach (1964)
			New Zealand	Parasite	S	DeBach (1964)
			Mauritius	Parasite	C	DeBach (1964)
			Kenya	Parasite	S	DeBach (1964)
			Madagascar	Parasite	S	DeBach (1964)
Hypera postica (Gyll.)	Alfalfa weevil	Alfalfa	California: San Francisco Bay Area San Joaquin Valley	Parasite	C	DeBach (1964)
					S	DeBach (1964)
			Parts of eastern U.S.	Parasite	S	van den Bosch (1971)
Oryctes tarandus Oliv.	Rhinoceros beetle	Sugarcane	Mauritius	Parasite	S	DeBach (1964)

(continued)

TABLE 9.1. (continued)

INSECTS

| Pest species | | Crop attacked | Place where controlled | Type of natural enemy | Degree of control | Reference |
Scientific name	Common name					
Promecotheca papuana Dziki	—	Coconut	New Britain	Parasite	S	DeBach (1964)
Promecotheca reichei Baly	Coconut leaf mining beetle	Coconut	Fiji	Parasite	C	DeBach (1964)
Rhabdoscelus obscurus (Bdv.)	Sugarcane weevil	Sugarcane	Hawaii	Parasite	S	DeBach (1964)
Diptera						
Ceratitis capitata (Wied.)	Mediterranean fruit fly	Many fruits	Hawaii	Parasite	S	Clausen *et al.* (1965)
Dacus dorsalis Hendel	Oriental fruitfly	Many fruits	Hawaii	Parasite	S	DeBach (1964)
Dasyneura pyri (Bouche)	Pear leaf midge	Pear	New Zealand	Parasite	S	DeBach (1964)
Orthoptera						
Oxya chinensis (Thumb.)	Chinese grasshopper	Sugarcane	Hawaii	Parasite	S	DeBach (1964)
Hymenoptera						
Cephus pygmaeus (L.)	European wheat stem sawfly	Wheat	Ontario (Canada)	Parasite	S	DeBach (1964)
	Spruce sawfly	Spruce	Canada	Parasite	S–C	DeBach (1964)

Pristiphora erichsonii (Htg.)	Larch sawfly	Larch	Canada	Parasite	S	DeBach (1964)

Hemiptera

| Nezara viridula (L.) | Green tomato bug or southern green stink bug | Vegetables | Australia | Parasite | S | DeBach (1964) |
| | | | Hawaii | Parasites | S–C | Davis (1967) |

WEEDS

Weed species		Place where controlled	Type of enemy	Degree control	Reference
Scientific name	Common name				
Alternanthera philoxeroides (mart.) Grisch.	Alligatorweed	Parts of southern Georgia and northern Florida	Leaf-feeding beetle	S	L. A. Andres (personal communication)
Clidemia hirta L.	Curse	Fiji	A leaf-feeding beetle	C	National Academy of Sciences (1968)
Cordia macrostachya (Jaquin) Roem. and Schult.	Black-sage	Mauritius	Leaf-and seed-feeders	S	National Academy of Sciences (1968)
Emex australis Steinh.	Spiny emex	Hawaii	Leaf-feeder	S	National Academy of Sciences (1968)
E. spinosa (L.) Campd.	Spiny emex	Hawaii	Leaf-feeder	S	National Academy of Sciences (1968)
Eupatorium adenophorum (Spreng.)	Pamakani	Hawaii	Gall-forming fly	C	National Academy of Sciences (1968)

(continued)

TABLE 9.1 (continued)

WEEDS

Weed species Scientific name	Common name	Place where controlled	Type of enemy	Degree control	Reference
Hypericum perforatum Linn.	Klamath weed or	California	Leaf-feeding beetles and a root borer	C	National Academy of Sciences (1968)
	St. Johnswort	Australia	A leaf-feeding beetle	S	National Academy of Sciences (1968)
		Chile	A leaf-feeding beetle	S	National Academy of Sciences (1968)
Lantana camara var. aculeata (L.) Moldenke	Lantana	Hawaii	A leaf-feeding bug and a leaf-feeding beetle	S	National Academy of Sciences (1968)
Opuntia. dilenii (Ker. Gawl.) Haw.	Prickly pear cactus	Ceylon	Pad-feeding mealybug	S	National Academy of Sciences (1968)
	Prickly pear	New Caledonia	Pad-boring caterpillar	S	National Academy of Sciences (1968)
O. imbricata Haw.) D.C.	Walkingstick cholla	Australia	Pad-feeding mealybug	C	National Academy of Sciences (1968)
O. inermis D.C.	Prickly pear cactus	Australia	Pad-boring caterpillar	C	National Academy of Sciences (1968)
O. megacantha Salm-Dyck	Mission prickly pear	Australia	Pad-feeding mealybug Pad-boring caterpillar	S	National Academy of Sciences (1968)
		S. India	Trunk borer Pad-feeding	S	National Academy of

			mealybug		Sciences (1968)
Opuntia stricta Haw.	Prickly pear cactus	Australia	Pad-boring caterpillar	S	National Academy of Sciences (1968)
O. triacantha Sweet	Prickly pear	West Indies	Pad-boring caterpillar	S	National Academy of Sciences (1968)
O. tuna Mill.	Prickly pear	Mauritius	Pad-feeding mealybug	S	National Academy of Sciences (1968)
			Pad-boring caterpillar		
O. vulgaris Mill.	Prickly pear cactus	Australia	Pad-feeding mealybug	S	National Academy of Sciences (1968)
Senecio jacobaea L.	Tansy ragwort	Limited areas of northwestern California and central Oregon	Leaf-feeding lepidopteran	S	Hawkes (1968)
Tribulus cistoides L.	Jamaica fever plant	Hawaii (Kauai)	Stem-boring beetle	C	National Academy of Sciences (1968)
T. terrestris L.	Puncturevine	Hawaii (Kauai)	Seed-feeding beetle	C	National Academy of Sciences (1968)

[a] A more complete list is given by Laing and Hamai (1976).
[b] Substantial.
[c] Complete.
[d] A native parasite.

Space limitation precludes analysis of each of the cases listed in Table 9.1 and in may cases, there simply are not enough data to permit meaningful analysis. Instead, we have elected to discuss in this chapter a limited number of cases that illustrate certain problems, approaches, or phenomena that have recurred in classical biological control programs (see Zwolfer *et al.*, 1976).

The Walnut Aphid

There are several noteworthy aspects to this highly successful program, namely: (1) it involved a pest that should have been controlled over half a century ago, (2) control was accomplished by a single natural enemy, the aphidiine wasp, *Trioxys pallidus* (Figure 9.1), (3) it involved deliberate use of an ecologically adapted strain of the wasp, and (4) it is a case in which initial failure was followed a decade later by virtually complete success.

Chromaphis juglandicola had been a serious pest of walnut in California since the early part of the current century. For decades, no thought at all was given to its biological control; instead, its annual outbreaks in thousands of acres of walnut were routinely treated with chemical insecticides. Initially, nicotine sulfate was the most widely used material; however, since the mid-1940s it has been supplanted by a variety of synthetic organochlorines and organophosphates. Chemical control of the aphid is costly, disturbs the walnut ecosystem, causes rapid resurgence of the pest, can create "secondary pest" problems, and is hazardous to warm-blooded animals, honeybees, and other pollinators. Furthermore, the aphid had developed resistance to a succession of insecticides used against it. One wonders then why biological control was not attempted sooner than it was. The answer simply seems to be that no one gave it serious thought. The aphid was in the groves virtually from the time that commercial walnut production was undertaken in California. Apparently most persons involved in walnut production and walnut pest control routinely attacked it with insecticides, giving little thought to its possible permanent suppression by biological control. It was no help either that for years (the woolly apple aphid case notwithstanding) there was a widely held belief that aphids were poor prospects for biological control, presumably because their high reproductive capacities and ability to develop at low temperatures gave them an insurmountable advantage over natural enemies.

The idea that aphids were not generally susceptible to classical biological control was completely dispelled by the striking success scored against the spotted alfalfa aphid in California in the middle 1950s (van den Bosch *et al.*, 1964). This success turned thoughts to other aphid species as possible targets of biological control, a particularly ripe field since virtually all of California's pest aphids are exotic species. Among these, the walnut aphid was selected as a prime target. There were several reasons for this: (1) in California it is monophagous and thus restricted to the walnut habitat, which eliminated

Figure 9.1. The walnut aphid and its parasite *Trioxys pallidus*. (A) *T. pallidus* searching among walnut aphids on the lower surface of a walnut leaf (photo by P. F. Daley). (B) A *T. pallidus* female ovipositing in an aphid nymph (photo by Jack K. Clark).

concern over the need for finding parasites adapted to both primary and alternative hosts plants and their habitats; (2) the walnut ecosystem is a stable, relatively long-lived system and thus favorable to the establishment of the interaction; (3) *C. juglandicola* was obviously an insect suitable for attack by an effective natural enemy in that it was generally and perennially epidemic, and its populations were free of significant attack by competing parasites, i.e., the parasite niche was vacant; and (4) related Callaphididae (including the spotted alfalfa aphid) were known to have important parasites in the Old World.

As we mentioned in Chapter 8, the initial importation of *Trioxys pallidus* from France in 1959 had only limited success in the milder coastal valleys of Southern California, but it failed to become established in the major walnut-producing areas in the hotter, drier interior valleys. Another strain of *Trioxys pallidus* was obtained from Iran in the late spring of 1958 and colonized in several areas of central California during the summer and autumn of that year. Results were prompt and spectacular. Even though only small colonizations were made at the most unfavorable time of year (midsummer), recoveries were made at all colonization sites. Furthermore, an autumn survey at one of the colonization sites showed that the parasite had effected a high level of parasitization and had spread at least a mile from the colonization focus. The wasp then survived the winter of 1968–1969 and continued to increase in abundance and to disperse. Additional colonizations were made in 1969, and establishment was again invariably effected. By the autumn of 1969, *T. pallidus* was widely established in the Great Central Valley and in several valleys peripheral to San Francisco Bay. Surveys conducted in the spring and early summer of 1970 showed that the parasite was having a heavy impact on *C. juglandicola* at a number of places. By the end of that year, it had virtually covered all of the walnut-growing areas of central and northern California, an area of perhaps 50,000 square miles (van den Bosch *et al.*, 1970). By the spring of 1971, *T. pallidus* had a generally crushing impact on the aphid. Surveys indicated that economically injurious infestations of the aphid were virtually nonexistent in the major production areas. In two monitored groves, parasitization of the fundatrix generation (aphids hatching from overwintering eggs) exceeded 90%. This degree of parasitization appeared to be general in commercial groves.

After ten years of vigorous activity, the Iranian *T. pallidus* continues to be a major population determinant of *C. juglandicola* populations. Already the wasp can be credited with almost total commercial control of this aphid.

The Winter Moth

The winter moth (*Operophtera brumata L.*) program is especially notable because the two parasites that effected essentially complete control of the winter moth in Nova Scotia and neighboring eastern provinces in Canada were not particularly prominent species in their area of indigeneity (Europe) (Embree, 1971). The important implication here is that it is a serious mistake to prejudge imported natural enemies on the basis of their performance in their native habitats.

This moth (Geometridae) is of European origin and its larvae feed on the foliage of hardwood species and thus pose a serious threat to forest, ornamental, and orchard trees. The pest was accidentally introduced into Nova Scotia some time in the 1930s, and by the middle 1950s it had covered approximately one-third of that province, occurring in moderate to severe infestations over several hundred square miles. In a span of ten years in just two counties, the moth destroyed 26,000 cords of oak wood valued at approximately $2 million. Furthermore, if its depredations had remained unchecked, it quite possibly would have wiped out all of the oak stands in Nova Scotia, caused severe damage to other forest species, shade, and orchard trees, and spread epidemically over much of Canada. The introduced parasites have unquestionably slowed the spread of the winter moth, and even if it moves into new areas, they appear capable of maintaining the pest at low epidemic levels.

Parasite introductions against *O. brumata* were initiated in 1954, and of six species colonized, two—the tachinid *Cyzenis albicans* (Fallen) and the ichneumonid *Agrypon flaveolatum* (Gravenhorst)—became established. By 1965, essentially complete biological control of *O. brumata* had been effected, with *C. albicans* being largely responsible for the pest's population collapse.

The winter moth program is one of the most meticulously analyzed biological control programs on record. The study is a particularly outstanding example of the utility of life table technique in demonstrating the crucial role of parasites in host population regulation. For example, in 1958 in a plot at Oak Hill, Nova Scotia, before parasitization had reached 10%, a total generation mortality rate of 96.5% was recorded, indicating an increasing population. But in 1960, when parasitization had reached 72%, the overall mortality percentage was 99.7% indicating a decreasing population. In each case, the trend indicated by the life table was confirmed by data taken the following year. Thus, in 1959 the population density showed a twofold increase while in 1961 it decreased by a factor of ten.

Analysis of 37 life tables compiled during the study permitted the development of a population model that showed that parasitism had indeed become the key factor in control of the winter moth. Thus, before the parasites rose to significant status, generation survival was correlated with egg and early instar survival. However, after parasitization increased beyond 19%, this relationship was completely reversed so that generation survival was correlated with late larval and pupal survival rather than with early instar survival. In addition, pupal survival was correlated with pupal parasitization.

Embree (1971), the principal investigator, quite forthrightly points out that "the success in this experiment can be described as mere good fortune," because the choice of the 6 parasites out of 63 species known to attack *O. brumata* for introduction into Nova Scotia was largely a matter of chance. Furthermore, the outstanding performance of *Cyzenis*, and to a lesser extent *Agrypon*, could not have been anticipated on the basis of their performance in Europe. On the other hand, the very careful studies of Embree and his group clearly reveal why the two parasites performed so effectively once they had become established in Canada. The key to this success is that the two species are compatible and complementary: one, *Cyzenis,* is effective at high host densities; the other, *Agrypon*, at low densities. Thus, as host abundance declines so does the efficacy of *Cyzenis*, but then at the lower densities *Agrypon* becomes relatively more efficient and in this way complements the former. This case, and the case of olive parlatoria scale in California cited later in this chapter, illustrate the conspicuous, beneficial, joint action that very commonly results from the introduction of more than a single "best" species.

The Oriental Fruit Fly

This program is particularly noteworthy because of the very striking successional sequence of the three key parasites involved. It provided an excellent opportunity for study of the competitive mechanisms of parasitoids within the host. The case is additionally significant because, as in the one involving the eucalyptus snout beetle (described later in this chapter), a parasite that attacks the earliest host developmental stage (egg) ultimately prevailed and effected a significant control.

Dacus dorsalis, as its common name implies, is an Asiatic species that invaded Hawaii, probably via military transport, sometime in the 1940s. Under the salubrious Hawaiian climate and with abundant host fruits available to it, the fly quickly erupted to enormous abundance and by the middle 1940s became a severe pest of a variety of fruits grown in the islands.

The severe local problem caused by *D. dorsalis* in Hawaii and the threat it posed to mainland agriculture engendered an intensive research, control, and quarantine program by a consortium of institutions and agencies consisting of the Hawaii Board of Forestry and Agriculture, the University of Hawaii, the U.S. Department of Agriculture, the California State Department of Agriculture, and the University of California Agricultural Experiment Station. Biological control was one of the earliest and most intensive control efforts undertaken by this group. Explorers were sent to virtually every tropical and subtropical area on earth, while reception, propagation, colonization, and evaluation teams were assembled in Hawaii. Actual parasite introduction was initiated in 1947–1948 by the Hawaii Board of Forestry and Agriculture, whose collectors shipped parasitized fruit fly material from the Philippines and Malaysia. Among the parasites obtained over the course of the program, three species —*Opius longicaudatus* (Ashmead), *O. vandenboschi* Fullaway and *O. oophilus* Fullaway,—played significant roles in the biological control of the pest, with the latter eventually dominating (Clausen *et al.*, 1965). By sheer chance the sequence of establishment of the three parasites was such that as time passed a pattern of succession unfolded in which the potentially least effective one (*O. longicaudatus)* first flourished and then was replaced by the next most promising species (*O. vandenboschi*), which in turn was displaced by the most promising one (*O. oophilus*).

The basis for this pattern lay in the ease of propagation of the parasites and thus in the number available for field colonization. *Opius longicaudatus* was the most easily propagated and hence the most heavily colonized, and the first to become widely established. This all came about because of the prevailing ignorance of the biologies of the three species at the time they were introduced. At that time, it was assumed that all three attacked medium-sized or large fruit fly larvae, and so larvae of these sizes were offered to the wasps for oviposition. But in actuality, only *O. longicaudatus* attacks larger fruit fly larvae, while *O. vandenboschi* attacks the first larval instar, and *O. oophilus* attacks the fruit fly egg.

It is really a wonder then that any *O. vandenboschi* and *O. oophilus* were propagated and colonized at all. In fact, establishment of the latter was simply sheer luck, because in being superficially similar to *O. vandenboschi* it was never recognized during the propagation and colonization phases of the program as a distinct species. It seems that quite by chance some of the infested fruits used in the parasite propagation program contained *D. dorsalis* eggs as well as partly developed larvae, and it was presumably on these that *O. oophilus*, mixed in with the *O. vandenboschi* stocks, reproduced itself.

The propagation and colonization history of *O. oophilus* can perhaps be considered an embarrassment to the entomologists involved in the Oriental fruit fly program, but it also teaches us a lesson about the great care that must be exercised in the introduction of natural enemies. In all fairness to those involved, it should be pointed out that there were great pressures to colonize the parasites during the crisis stage of the oriental fruit fly program. Over the course of 4 years, 14 explorers were in the field, and they shipped 4.5 million fruit fly puparia of more than 60 species to the Honolulu quarantine laboratory. These yielded over 20 species of parasites plus several predators. Ultimately, a total of 29 parasite and predator species were released in the islands. During the critical years 1947–1953, 1.1 million parasite and predator adults were produced by the insectary teams, and the accidental propagation and release of *O. oophilus* is thus more understandable. Fortunately, no undesirable species sifted through the screen, but we should be sufficiently chastened by the *O. oophilus* incident to make sure that no such happening recurs.

The imported parasites of *D. dorsalis,* especially *O. oophilus,* have had a striking impact on their host. All three parasite species were first liberated in the summer of 1948, and *O. longicaudatus* was recovered on the island of Oahu in October of that year, while *O. vandenboschi* and *O. oophilus* were first recovered about two months later. *Opius longicaudatus* increased rapidly, parasitizing up to 30% of the fruit fly larvae in wild guava by January 1949. This level of parasitization was sustained for several months thereafter. But then *O. vandenboschi* increased in abundance, and by October 1949 it had attained higher levels of parasitization than its predecessor. With the increase in importance of *O. vandenboschi, O. longicaudatus* rapidly faded to very low abundance. *O. oophilus* began to assert itself in the summer of 1950, and by the end of that year it was the overwhelmingly dominant parasite of *D. dorsalis*. In fact, its dominance is so great that over most of the Hawaiian Islands *O. longicaudatus* and *O. vandenboschi* have been relegated to the status of rare species, being in effect displaced.

O. oophilus has effected substantial biological control of *D. dorsalis.* Today the population level of the fly is strikingly lower than what it was before the parasites were introduced, and many crops, including avocado, banana, papaya, loquat, peach, and persimmon, are no longer seriously affected by it. Even in mango, a favorite host, infestation levels seldom exceed 10%, and this permits economical supplementary control with compatible sprays.

The dominance of *O. vandenboschi* and *O. oophilus* over *O. longicaudatus* derives from their earlier hatching in hosts and the resultant inhibition of the eggs and larvae of *O. longicaudatus* that may occur in the same hosts. Furthermore, in larvae simultaneously parasitized by *O. oophilus* and *O. vandenboschi,* the former invariable prevents development of the latter. Thus, with the very high levels of parasitization attained by *O. oophilus* there is virtually no chance for *O. longicaudatus* and *O. vandenboschi* to parasitize successfully a significant proportion of the fruit fly population. This is the secret to the success of *O. oophilus*.

The Imported Cabbageworm

The biological control of the imported cabbageworm, *Pieris rapae*, is presented to illustrate two points: (1) that a lack of knowledge of the host range of a parasite and factors determining that range can lead, after much effort, to little benefit; and (2) that it is valuable to reassess partially successful biological control projects in the light of our present awareness of the critical importance of host specificity and host suitability in the effectiveness of imported parasite species (or intraspecific strains).

The imported cabbageworm is one of the principal pests of cole crops (cabbage, brussels sprouts, collards, kale, broccoli). This pest, long established in North America, originated in Europe where it coexists with another cabbage-attacking butterfly, the European cabbageworm, *P. brassicae* (L.)

In Europe, larvae of both species are attacked by the gregarious, endoparasitic braconid, *Apanteles glomeratus*. It will be recalled from our chapter on the history of biological control that *A. glomeratus* was the first parasitic insect to be recorded in the literature, when in the early seventeenth century it was described emerging from larvae of *P. rapae*. This same braconid also bears the distinction of being the first parasite to be introduced (by C. V. Riley) into North America (1883) from another country (England) for the biological control of an insect pest (against *P. rapae*).

Numerous publications and reviews of biological control in the past 20–30 years refer to this biological control effort, stressing the historical role of the parasite, and generally referring to the effort as partially to moderately effective. However, up to and including the early 1960s, pesticide control of the imported cabbageworm was usually considered necessary in order to produce a commercially acceptable crop. In spite of its historical interest, *A. glomeratus* cannot really be credited with providing effective control of *P. rapae*.

However, our story is not ended. In fact, an important bit of detective work now unfolds. In 1956 a German expert in biological control, Hans Blunck, surveyed certain parts of the United States and Canada for natural enemies of the imported cabbageworm. He observed *A. glomeratus*, among other natural enemies, to be fairly generally distributed in the areas he searched, attaining levels of parasitism of up to 50% in certain California cole crop fields. However, he was surprised over his failure to discover another *Apanteles* species very common on *P. rapae* in Europe, namely, *A. rubecula* Marshall. Blunck (1957) noted that in Germany *A. glomeratus* seldom parasitized *P. rapae*. Rather, this host is very frequently attacked instead by *A. rubecula*. He indicated that *A. glomeratus*, on the other hand, is the most common parasite of the related host, *P. brassicae*. Richards (1940) earlier reported similar results in England.

Then, in the period 1963–1965, Canada Department of Agriculture entomologist A. T. S. Wilkinson, in making a survey of parasites of the imported cabbageworm in the south coastal areas of British Columbia, discovered the presence for the first time in North America of *A. rubecula* (Wilkinson, 1966). It was found to be well established, the most abundant of all parasites (others were two tachinids, but not *A. glomeratus*), and the most widespread. Wilkinson was unable to explain the origin of this parasite, but we can presume that it came to British Columbia accidentally, perhaps as cocooned pupae or as larvae in host caterpillars carried on cabbage conveyed in commercial aircraft or in maritime ships' stores.

Stressing again that *A. rubecula* is almost specific to *P. rapae*, Wilkinson recorded it to parasitize the host up to levels of 50%. It showed itself to be a solitary endoparasite, emerging from fourth-stage host larvae rather than fifth-stage larvae, as in the case of the gregarious *A. glomeratus*. In 1967, after initiating a new project on the biological control of cole crop pests in Missouri, U.S. Department of Agriculture entomologists theorized that *A. rubecula* might very well be an important importation candidate for control of *P. rapae* (Puttler *et al.*, 1970). At that time, *P. rapae* in Missouri required pesticide applications to keep it in check, even though *A. glomeratus* was present there. Puttler and his associates (1970) reported some additional important clues regarding the suitability of *P. rapae* as a host for *A. glomeratus*. First they cited Boese (1936), who reported that eggs of *A. glomeratus* are sometimes encapsulated after being oviposited into *P. rapae* larvae. Puttler and colleagues also found this to be the case in Missouri whenever host larvae larger than first instar were attacked. These findings, coupled with the evidence that *A. rubecula* eggs are not encapsulated in this host and that it is almost specific to *P. rapae* (Richards, 1940), caused them to conclude that *A. glomeratus* is not closely adapted to this host species but that *A. rubecula* is.

Several colonies of *A. rubecula* were imported from British Columbia to Missouri and colonized on *P. rapae* in 1967. At that same time, the egg parasite *Trichogramma evanescens* Westwood, known to be naturally adapted to *P. rapae,* was imported from eastern Europe.

U. S. Department of Agriculture entomologist F. D. Parker noted that at the beginning of each growing season, cabbageworm populations occurred in very low numbers and were essentially unparasitized but then rapidly exploded to damaging levels, after which time parasitization rose to substantial levels. He decided to release large numbers of the newly imported natural enemies at the beginning of the season so as to promote a more rapid and effective buildup of the control agents (Parker, 1971; Parker *et al*, 1971).

This attempt was only partially successful; parasitization did rise more rapidly but still not sufficiently to prevent damage. The host was so rare at the start of the season that the parasite had great difficulty in finding them before they "escaped." Parker reasoned that if there were more hosts during late spring, the parasites would be able to increase more effectively.

Release in the late spring of both *P. rapae* eggs and parasite adults (both species) provided just the right combination of numbers. The host started to build up as usual, but now the natural enemies increased even more rapidly. In spite of the addition of the very pest being controlled, the cabbageworm population was curtailed before damage occurred in the late summer and fall.

Thus, through the imported cabbageworm had been the subject of a biological control effort since 1883, efforts to obtain the properly adapted natural enemies associated with it in its native home were only attempted in recent years. Information on the existence of the "true" parasite of *P. rapae* together with evidence of the maladaptedness of the original import, *A. glomeratus*, had been available for more than 25 years (Boese, 1936; Richards, 1940). The lesson this holds is clear: Each case of biological control must be carried out with great attention to all the subtle details of host and natural-enemy distribution, parasite adaptedness, and seasonal host–parasite synchrony before success can be anticipated with confidence. Such efforts are never likely to be made unless an enthusiastic, well-trained, and clearly identified group of specialists are charged with the full direction of such programs and are given adequate funds to explore all possibilities. This has not commonly been the case for much of the biological control work done throughout the world.

The Eucalyptus Snout Beetle in South Africa

This project constitutes a classic example of the control of an invader pest by means of a natural enemy imported from the pest's home grounds. It further provides an example of control by an egg parasite, a type of natural enemy not previously considered to be particularly effective.

The scene of this accomplishment is South Africa, the protected plant, the eucalyptus tree, a native of Australia. As in the southwestern U. S., in South Africa, eucalyptus was imported from Australia during the nineteenth century to serve as a rapid-growing, potential hardwood timber source. As in the United States, the establishment of eucalyptus was spread throughout those parts of South Africa suitable to it, and plantation stands and windbreaks thrived.

Then in 1916, a pest of eucalyptus, identified as *Gonipterus scutellatus* (Gyllenhal), was discovered near Cape Town. In the following years, the weevil, whose adults and larvae feed on leaves and new shoots, spread rapidly, and it soon became apparent that "the snout beetle was threatening the entire eucaluptus-growing industry of the country" (Tooke, 1953).

The expert identification of the pest provided the evidence of its origin. All of the members of the curculionid genus *Gonipterus* originate in Australia. *G. scutellatus* may be said to have coevolved with *Eucalyptus*.

In 1925, South African entomologists initiated a biological control program against the pest. For several years prior to this time it was realized that in the native home of the snout beetle, Tasmania and Australia, it was not considered a pest and was not prominent. In fact, its biology and ecology were little known. The entomologists concluded from this state of affairs that the snout beetle was not a damaging pest in its native home, and that this might be due to the presence in Australia of effective natural enemies. So, in 1926, entomologist, F. G. C. Tooke was sent to Adelaide, South Australia, to search for natural enemies and to observe the economic status of the snout beetle, the species of Eucalyptus attacked, and the climatic conditions prevailing in the infested areas.

Since Australian entomologists were uncertain as to where snout beetle infestations might be found, Tooke proceeded to host tree stands to search foliage for signs of weevil damage. Within a month he found egg masses of the pest on eucalyptus leaves in southeastern coastal South Australia. Dissection showed that some of the eggs contained in the egg capsules were blackened, a sign of egg parasitism that proved to be due to a mymarid wasp, later named *Anaphoidea nitens* Girault (=*Patasson nitens*). He collected and shipped some of these egg capsules to South Africa; later several tachinid parasites were also discovered, and they too were shipped to South Africa.

Tooke considered the egg parasite to be quite promising, as indeed it later proved to be. Collections showed it to be the most abundant natural enemy of the pest. Early-season egg parasitization rates reached nearly 70%. Many rearings disclosed no hyperparasites. He did encounter a shipping problem, however, as the developmental period of the egg parasite was found to be shorter than the 3-week duration of the sea voyage from Adelaide to Durban, South Africa. This required cold-storage shipment, necessitating a study of the consequences of cold exposures to the parasite.

The first shipments of *A. nitens* were received in good condition in South Africa in 1926. Colonization was effected promptly, and successful establishment was observed in 1927. During the next 4 years, a large-scale insectary propagation program aided in the dissemination of the parasite. Rapid natural dispersal also occurred.

The results of this importation were dramatic. Complete control occurred within 3 years in climatically suitable areas of South Africa. By 1941 economic control was produced on all but one eucalyptus species, and that in the climatically severe regions.

The effectiveness of the egg parasite was unexpected. Conventional belief among biological control workers was that egg parasites were not promising control agents. This attitude was based partly on the rather poor showing attained up to that time from the extensive work with *Trichogramma* egg parasites and the fact that no previously successful biological control program was due to an egg parasite. It was also felt that most pests susceptible to biological control were heavily attacked in the larval stages by parasites, so that any mortality produced by egg parasitism would have occurred in any case later in the life cycle. Egg parasites were thought less dependable, too, by virtue of their being considered more nonspecific and polyphagous in habit.

The success of the eucalyptus snout beetle program proved that an egg parasite can be an effective control agent. It also confirmed what a number of other successful biological control cases have since demonstrated: that effective control could be accomplished by just one parasite species. This, too, was not anticipated, since the biological control theory at the time held that a sequence of natural enemies attacking the successive stages of a pest during its life cycle was more promising of control than merely a single enemy.

The Olive Scale in California

This project illustrates two important elements of biological control. One of these is the existence in a genus of sibling groups of natural-enemy species that may masquerade for years as a single species, some of which turn out to be ineffective, while others prove to be effective. The problem is to recognize and perhaps capitalize on this situation. The other unique element in this project is the complementary action of two parasites, both attacking the same host population.

Olive scale, *Parlatoria oleae*, was the most serious pest of olive in California from soon after the time of its discovery in 1934 until after a biological control campaign was begun in 1948 (Huffaker *et al.*, 1962). Besides olive, this polyphagous scale attacks some 200 host plants, including peach, apricot, plum, almond, and numerous ornamentals. The pest attacks not only twigs and leaves but also fruits. During the period before biological control was accomplished, commercially acceptable crops of olives could not be produced without use of pesticide applications.

Parlatoria oleae is believed native to northern India and Pakistan, though it has been known from the Mediterranean region for centuries. Initial searches for natural enemies were made in the Mediterranean countries, since the solitary ectoparasite *Aphytis maculicornis* (Figure 9.2) had been known to be associated with olive scale in Italy since 1911. First importations of *A. maculicornis* were from Egypt. This strain became established by 1949, though parasitization rates by it were never substantial. In subsequent years, additional imports of the *A. maculicornis* complex of strains and sibling species were made from India, Iran, and Spain. Careful studies of life histories of these different imports established that each could be classed as a "strain" because of certain biological differences (Hafez and Doutt, 1954) or sibling species, as Rosen and DeBach (1976) later classified the Iranian form as *A. paramaculicornis*. The original Egyptian stock exhibited a thelytokous mode of reproduction,—no males known, females produce all female offspring by parthenogenesis. The Spanish strain was deuterotokous —virgin females producing mostly female offspring (but with some males) parthenogenetically. The Iranian and Indian stocks were both arrhenotokous — offspring of both sexes being produced only after mating, males only being

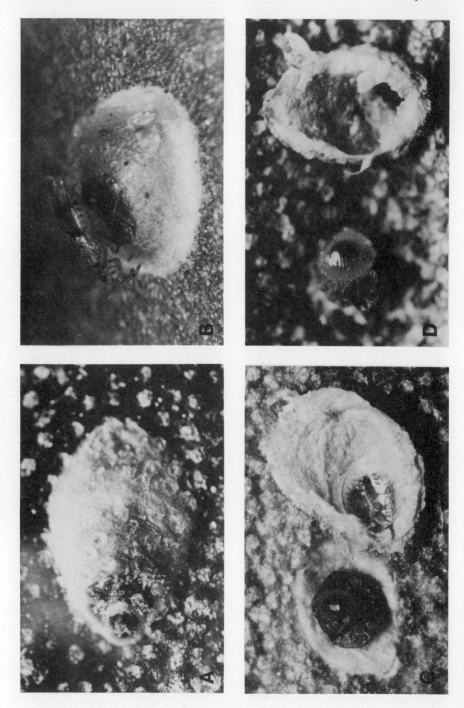

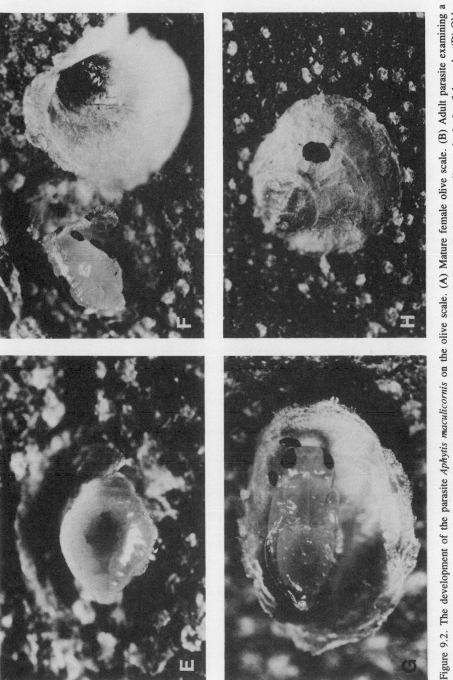

Figure 9.2. The development of the parasite *Aphytis maculicornis* on the olive scale. (A) Mature female olive scale. (B) Adult parasite examining a mature scale preparatory to oviposition. (C) Scale covering removed to show young parasite developing externally on the body of the scale. (D) Older parasite larva as in (B). (E) Mature parasite larva as in (B). (F) Early parasite pupa. (G) Parasite pupa showing eye coloration and black meconial pellets. (H) Exit hole left by the adult parasite (photos by Frank E. Skinner).

produced parthenogenetically by unmated females. Differences also existed in the rates of development of these strains.

Following colonizations of the Egyptian strain of *A. maculicornis* in 1948–1949, the Iranian, Indian, and Spanish stocks of *Aphytis* were released in 1952–1953. Only the Iranian colonizations, i.e., *A. paramaculicornis*, increased in numbers to high levels. The assumption made later was that in all probability only this stock possessed the required adaptation to persist at moderate levels in California's environments.

During the period 1952–1960, millions of the Persian *Aphytis* were colonized throughout much of California. Establishment was soon verified in many localities. But the degree of biological control was variable and in some cases not commercially adequate. The main reason for this is the harm caused to the parasite by the intensive summer heat and low humidity that occurs in the commercial olive-growing areas of California.

Nevertheless, in many olive groves very satisfactory commercial control of the scale was provided by *A. paramaculicornis*. Furthermore, subsequent to the establishment of this parasite, there followed a marked general decline in the pest on essentially all of its other host plants.

During the importation efforts of 1951–1952, which led to the discovery and successful establishment of the Iranian species of *A. paramaculicornis*, several other species of parasitic Hymenoptera were reared from imported olive scale samples in the quarantine laboratory. Several of these proved to be hyperparasitic and hence were exterminated. One turned out to be an endoparasitic species of *Coccophagoides*, which because of complications in its life cycle, proved difficult to rear in the laboratory. The culture of this parasite was subsequently lost (Broodryk and Doutt, 1966).

With *A. paramaculicornis* (=*maculicornis* in literature) providing only partially effective biological control, attempts were again made to acquire this new species of *Coccophagoides*. The import came from India and Pakistan in 1951–1952. In 1957, explorer–entomologist Paul Debach, while searching for a parasite of the citrus red scale in the Near and Far East, rediscovered the *Coccophagoides* sp. on olive scale infesting apple in Pakistan. Shipments of material to California led to the successful reacquisition of the new species in the quarantine laboratory. It was subsequently named *Coccophagoides utilis* Doutt, (Figure 9.3).

The reason *C. utilis* was so difficult to rear was because males are produced only as hyperparasitic ectoparasites on immature female parasites of their own species in the same scale, while females are produced in the normal way as endoparasites on olive scale. This process of autoparasitism (adelphoparasitism of males on females of the same species) is common in the group of genera to which *Coccophagoides* belongs.

To rear this parasite, and later to mass-culture it for biological control colonizations, careful adjustments of the culture must be arranged. This

Figure 9.3. Photos of the olive scale and the parasite *Coccophagoides utilis*. (A) Female *C. utilis* ovipositing on a female olive scale. (B) Dead olive scales from which *C. utilis* have emerged (photos by Jack K. Clark).

species is arrhenotokous—mating is required to produce females. Mated females must be used to attack healthy olive scales; some of the latter must be separated, held until the immature parasite reaches the prepupal or pupal stage, and then exposed to unmated females for the production of male progeny. The remainder of the scales parasitized by mated females are then held for emergence of female progeny. Males are eventually recovered from parasite pupae (their own sisters or cousins) contained in the dead host scales. Only part of the newly emerged females can be exposed to males for mating; the remainder must be kept unmated to produce more males. This unusual biology can cause complications in culture and establishment, particularly in a program where there is an attempt to rear the thousands of parasites needed for colonization work.

After a culture technique for *C. utilis* was devised, the parasite was colonized at several locations in California during the period 1957–1958. Establishment was proven to have occurred in 1961, when the parasite was recovered from two of the original release sites. It was found to occur in association with the previously established *A. maculicornis*. Subsequent studies disclose that rather than competing with each other to the detriment of the combined control effect, these two parasites supplemented one another. This supplementing effect was of such a nature as to result in excellent biological control of the scale, considerably better and more reliably than the better species, *A. paramaculicornis*, could achieve alone and as good as the control provided by the "best" insecticide.

While *A. paramaculicornis* was shown to be adversely affected by summer heat and low humidity, *C. utilis* on the other hand manages to survive the summers without undue losses. Being an excellent searcher of scales, it manages to produce levels of parasitization reaching 60%, even at low scale densities. So, while *A. paramaculicornis* thrives during the spring though suffering in the summers, *C. utilis* presists with little mortality during summer and supplements the former somewhat during the spring.

This project thus provides interesting lessons, namely: (1) that it is important to discover the proper natural-enemy strain or close sibling species; (2) that two parasites can be better than one, that multiple introductions can definitely lead to improved results over single species introductions; and (3) that potentially competitive parasites can coexist, provided resource use differs or external factors such as climate shift the competitive advantage from one to the other and back again.

St. Johnswort or Klamath Weed

This project has been one of the most successful cases of weed control through use of plant-feeding insects, and it provides a very good example of how such a project should be carried out. It is also a very striking example of how important a herbivorous insect can be in influencing the ecology of a plant.

Klamath weed, *Hypericum perforatum*, known in other countries as St. Johnswort and a perennial native of Europe and Asia, has invaded many semiarid to subhumid temperature regions of the world, notably Australia, New Zealand, Argentina, Chile, South Africa, and North America. It was first found in northern California near the Klamath River (hence its American name) in the early 1900s. During the subsequent years, it spread throughout the low-elevation valleys, canyons, and coastal plains of California, occupying valuable rangelands, until by the mid-1940s it covered more than 2 million acres in California alone (Holloway and Huffaker, 1952).

The weed is very aggressive in the dry western United States, and as it expanded its distribution, it displaced native and introduced forage plants of prime value to livestock. Besides displacing more suitable forage plants, Klamath weed is also poisonous to livestock. While chemical and cultural controls were possible as means of combatting the weed, such measures were far too costly—the weed covered too great an area, and much of it was inaccessible.

In Australia, a biological control program against *H. perforatum* was begun in 1920, with first attempts to obtain and test insect-feeders of the weed concentrated in England. Lack of success caused a shift in search to southern France in 1935. There, three species of insects, one seen in England, were found attacking the plant, and exhaustive testing to determine the host specificity of the insects was commenced.

One of the major tasks in a project on biological control of weeds is the determination, beyond any reasonable scientific doubt, that any candidate insect to be used in such a control program will not shift its attention from the target weed to some other plant, particularly one of economic importance such as a crop plant. Experience with insect feeding habits indicates that many phytophagous insect species are polyphagous, while many others are oligophagous, attacking several closely related plants. For weed control purposes, an insect must be monophagous, or at least confine its attack to closely related plant species that are themselves of no economic or ecological importance.

Three species of beetles from France, having passed the rigorous feeding tests beforehand, were sent to Australia and colonized in the late 1930s. Eight years later, observations showed that two of the beetles were established and control results were promising.

In 1944 a similar biological control project was begun in California. Because of World War II, which prevented entomological work in France, the three beetles—*Chrysolina hyperici* (Forster) and *C. quadrigemina* (Suffrian), which are leaf feeders, and *Agrilus hyperici* (Creutzer), which is a root borer—were obtained from Australia, and, after some additional host-specificity testing, the *Chrysolina* were initially colonized in California in 1945–1946.

The beetles have but one generation per year, so that the results of a colonization could not be evaluated very rapidly. Nevertheless, within 3 years it was found that the two leaf-beetle species (*Chrysolina*) not only had become established, but were flourishing. The target weed was being severely defoliated at the release sites, so much so that within 3 years the hardy perennials were being killed back (see Figure 9.4).

The beetles spread slowly as long as Klamath weed was abundant, even though they were found to be capable of flights of as much as two to three miles radially from a point source. But by artificial distributions of field-collected beetles, the entire infested area in California was colonized by 1950. Other western states also acquired the beetles.

The final results, attained by 1956 and still prevailing today, were striking. The dense, extensive stands of the weed were devastated. Only in certain shady sites in narrow canyons and under trees was the weed able to hold out initially, and even here it has now further declined, partly due to an assist by the root borer *A. hyperici*. Cattle growers were enthusiastic, as native forage plants (bunchgrasses), Mediterranean grasses, and naturalized clovers reoccupied the range (Huffaker and Kennett, 1959). The control program was a resounding success, so much so that the grateful ranchers erected a monument to the successful beetles.

Ironically, the results of the control campaign against *H. perforatum* in Australia were nowhere as decisive. While suppression of the weed by the two *Chrysolina* spp. plus at least one other insect imported into Australia was noticeable, it was not as dramatic nor as complete as in the western United States (Huffaker, 1966).

Attempts were also made in 1951–1952 to establish the two *Chrysolina* species on Klamath weed in British Columbia, Canada (Smith, 1958). Colonization attempts gave variable results, with quick solid establishment occurring in only a few sites. Curiously, probably as a consequence of climatic stress, it was not *C. quadrigemina* but *C. hyperici* that was the better performer at least initially, so far as relative numbers go. However, weed suppression was disappointing and biological control of *H. perforatum* in British Columbia remains only partially successful; the work is continuing.

One of the major objections to biological control of weeds by use of phytophagous organisms raised by agricultural scientists and public authorities is the fear that insects imported to control a plant by feeding on it will shift their attention to other plants. The California experience provides evidence as to the stability of the plant–herbivore relation in insects used in weed control work. This holds true despite the fact that some question has recently been raised that the Klamath weed beetles also attack *Hypericum calycinum* L. in the coastal region of California, which in recent years has been introduced extensively as an ornamental species. Here we see a minor conflict develop, not because it was unanticipated, but because a closely related host species (tested and known to have some host potential) was selected as an ornamen-

Figure 9.4. Biological control of Klamath weed. (A) Rangeland in Humbolt County, California, three years after introduction of beetles (photo by J. K. Holloway). (B) Same area two years later showing the complete destruction of the weed and the repopulation of valuable forage plants (photo by J. Hamai).

tal. Small anticipated changes in feeding habits are likely to occur as the employed insects accomplish a substantial destruction of the target pest species and then come under severe starvation pressures. Thus we see that the results of very stringent feeding tests conducted on the candidate species by researchers before colonization pinpointed the extent of the potential problem (i.e. no plants other than *Hypericum* have been attacked and damaged in any way), indicating that the biological control of weeds, if properly conducted, is an entirely safe and effective procedure.

References

Blunck, H. 1957. *Pieris rapae* (L.) its parasites and predators in Canada and United States. *J. Econ. Entomol.* **50**:835–836.

Boese, G. 1936. Der Einfluss tierscher Parasiten auf den Organismus der Insecken. *Z. Parasitenk.* **8**:253–284.

Broodryk, S. W., and R. L. Doutt. 1966. Studies of two parasites of olive scale *Parlatoria oleae* (Colvee) in California. 2. The biology of *Coccophagoides utilis* Doutt (Hymenoptera, Aphelinidae). *Hilgardia* **37**:233–254.

Clausen, C. P., D. W. Clancy, and Q. C. Chock. 1965. Biological control of the Oriental fruit fly (*Dacus dorsalis* Hendel) and other fruit flies in Hawaii. U. S. Dept. Agric. Tech. Bull. 1322. 102pp.

Davis, C. J. 1967. Progress in the biological control of the southern green stink bug *Nezara viridula* variety *smaragdula* (Fabricius) in Hawaii (Heteroptera; Pentatomidae). *Mushi* **39**:9–16.

DeBach, P. 1964. Successes, trends, and future possibilities. In: P. DeBach (ed.) *Biological Control of Insect Pests and Weeds*, Chap. 24. Chapman & Hall: London. pp. 673–713.

DeBach, P., and L. C. Argyriou. 1967. The colonization and success in Greece of some imported *Aphytis* spp. (Hym. Aphelinidae) parasitic on citrus scales (Hom. Diaspididae). *Entomophaga* **12**:325–342.

Embree, D. G. 1971. The biological control of the winter moth in eastern Canada by introduced parasites. In: C. B. Huffaker (ed.) *Biological Control* Chap. 9. pp. 217–268. Plenum Press: New York. 511 pp.

Hafez, M., and R.L. Doutt. 1954. Biological evidence of sibling species in *Aphytis maculicornis* (Masi) (Hymenoptera, Aphelinidae). *Can. Entomol.* **86**:90–96.

Hawkes, R. B. 1968. The cinnabar moth, *Tyria jacobaeae,* for control of tansy ragwort. *J. Econ. Entomol.* **61**:499–501.

Holloway, J. K., and C. B. Huffaker. 1952. Insects to control a weed. In: *Insects*, Yearbook of Agriculture for 1952, pp. 135–140. U. S. Govt. Printing Office: Washington, D. C.

Huffaker, C. B. 1966. A comparison of the status of biological control of St. Johnswort in California and Australia. In : *Natural Enemies in the Pacific Area* Eleventh Pac. Sci. Cong., Tokyo. pp. 51–73.

Huffaker, C. B., and C. E. Kennett. 1959. A ten-year study of vegetational changes associated with biological control of Klamath weed. *J. Range Manage.* **12**:69–82.

Huffaker, C. B., and C. E. Kennett. 1966. Studies of two parasites of olive scale, *Parlatoria oleae* (Colvee). 4. Biological control of *Parlatoria oleae* (Colvee) through the compensatory action of two introduced parasites. *Hilgardia* **37**:283–335.

Huffaker, C. B., C. E. Kennett, and G. L. Finney. 1962. Biological control of olive scale, *Parlatoria oleae* (Colvee) in California by imported *Aphytis maculicornis* (Masi) (Hymenoptera, Aphelinidae). *Hilgardia* **32**:541–636.

Kobakhidze, D. N. 1965. Some results and prospects of the utilization of beneficial entomophagous insects in the control of insects in Georgian S.S.R. (USSR). *Entomophaga* **10**:323–330.

Laing, J. E., and J. Hamai. 1976. Biological control of insect pests and weeds by imported parasites, predators, and pathogens. In: C. B. Huffaker and P. S. Messenger (eds.) *Theory and Practice of Biological Control.* Academic Press: New York. pp. 685–693.

Mathys, G., and E. Guignard. 1965. Etude de l'efficacité de *Prospaltella perniciosi* Tow. en Suisse parasite du pou de San Jose. *Entomophaga* **10**:193–220.

National Academy of Sciences. 1968. The biological control of weeds. In: *Principles of Plant and Animal Pest Control.* Vol. II, Chap. 6. Nat. Acad. Sci. Publ. 1597. pp. 86–119.

Parker, F. D. 1971. Management of pest populations by manipulating densities of both hosts and parasites through periodic releases. In: C. B. Huffaker (ed.) *Biological Control,* Chap. 16. Plenum Press: New York. pp. 365–376.

Parker, F. D., F. R. Lawson, and R. E. Pennell. 1971. Suppression of *Pieris rapae* using a new control system: Mass releases of both the pest and its parasites. *J. Econ. Entomol.* **64**:721–735.

Puttler, B., F. D. Parker, R. E. Pennell, and S. E. Thewke. 1970. Introduction of *Apanteles rubecula* into the United States as a parasite of the imported cabbageworm. *J. Econ. Entomol.* **63**:304–305.

Richards, O. W. 1940. The biology of the small white butterfly (*Pieris rapae*), with special reference to the factors controlling its abundance. *J. Anim. Ecol.* **9**:243–288.

Rosen, D., and P. DeBach. 1976. Biosystematic studies on the species of *Aphytis* (Hymenoptera: Aphelinidae) *Mushi* **49**:1–17.

Shuster, M. F., J. C. Boling, and J. J. Marony, Jr. 1971. Biological control of Rhodesgrass scale by airplane releases of an introduced parasite of limited dispersing ability. In: C. B. Huffaker (ed.) *Biological Control,* Chap. 10. Plenum Press: New York. pp. 227–250.

Smith, J. M. 1958. Biological control of Klamath weed, *Hypericum perforatum* L. in British Columbia. *Proc. 10th Int. Congress Ent. Montreal* (1956) **4**:561–565.

Tooke, F. G. C. 1953. The eucalyptus snout-beetle, *Gonipterus scutellatus* Gyll. A study of its ecology and control by biological means. Union S. Africa, Dept. Agric. Entomol. Mem. 3. 283 pp.

van den Bosch, R. 1971. Biological control of insects. *Annu. Rev. Ecol. Syst.* **2**:45–66.

van den Bosch, R., E. I. Schlinger, J. C. Hall, and B. Puttler. 1964. Studies on succession, distribution, and phenology of imported parasites of *Therioaphis trifolii* (Monell) in Southern California. *Ecology* **45**:602–621.

van den Bosch, R., B. D. Frazer, C. S. Davis, P. S. Messenger, and R. Hom. 1970. *Trioxys pallidus*—An effective new walnut aphid parasite from Iran. *Calif. Agric.* **28**:8–10.

Wilkinson, A. T. S. 1966. *Apanteles rubecula* Marsh and other parasites of *Pieris rapae* in British Columbia. *J. Econ. Entomol.* **59**:1012–1018.

Zwolfer, H., M. A. Ghani, and V. P. Rao. 1976. Foreign exploration and importation of natural enemies. In: C. B. Huffaker and P. S. Messenger (eds.) *Theory and Practice of Biological Control.* Academic Press: New York. pp. 189–205.

10

Naturally Occurring Biological Control and Integrated Control

Faunistic surveys of agricultural, sylvan, or natural, undisturbed environments will disclose large numbers of herbivorous insect species that are of insignificant abundance, causing little or no harm to the plants growing in such habitats. Many of these insects are kept in check by native natural enemies. We describe this situation as *naturally occurring biological control.*

Such natural enemies as those described above are often overlooked in conventional biological control efforts by man because they and their hosts occur in low numbers and are not often observed in the environments concerned. Their host species are rarely considered pests and hence have usually been ignored by agricultural growers.

In the years following 1945, the new powerful, synthetic organic insecticides appeared and made insect suppression so easy and effective that virtually every other method of insect control was dropped in favor of these miracle chemicals; they became our insect control "crutch." But in neglecting these investigations and the development of alternative controls, we rendered ourselves vulnerable to insect depredation in the event that the chemicals failed (van den Bosch, 1978). Very few imagined that this would ever come about, although in 1940 H. S. Smith of California had widely warned of such an occurrence. But today it is happening, and because we do not have pest control systems that embrace alternative methods, major disasters threaten in may areas. The signs are manifold: In northeastern Mexico, the cotton industry was destroyed by the tobacco budworm, *Heliothis virescens* Fabricius, an insect that can no longer be killed by insecticides (Adkisson, 1971); in Central America, the human population, virtually malaria-free for more than a decade, is now threatened with a devastating epidemic as the vector mosquito, *Anopheles albimanus* Wiedemann, verges on total resistance to available insecticides (Georghiou, 1971); in California's Great Central Valley, similar resistance has developed in the encephalitis vector, *Culex tarsalis* Coquillet (Georghiou, 1971). In the wake of insecticide usage, spider mites have leapt to the forefront as crop pests and threaten economic disaster in a variety of commodities the world over (Huffaker *et al.*, 1970; McMurtry *et al.*, 1970). Today there are more insect species of pest status than ever before, insect control costs have mounted strikingly, and insecticide threat to human health and its environmental pollution has become a problem of global proportions. In California alone, the majority of all insect pests are resistant to one or more of these chemicals (Luck *et al.*, 1977) and the situation is not likely to be much better in other areas of the world.

Effects of Insecticides

A number of factors have contributed to this alarming situation, but insecticide disruption of naturally occurring biological control lies very much at the heart of the matter. In this connection, it is now glaringly apparent that in the mid-1940s, when DDT burst onto the scene, there was an appalling lack of awareness of the role of naturally occurring biological control in insect population regulation. Quite evidently, some entomologists knew that natural enemies could control insect populations, for, as described here, a number of exotic pests had been controlled by introduced natural enemies before 1940. But many entomologists in pest control gave little thought to the possibility that the same process of population regulation could affect endemic pests— that native species might be held at innocuous levels by natural enemies. In other words, a sort of narrow-minded view appears to have developed concerning the role of natural enemies in pest control. On the one hand, the deliberate reassociation of invading insects and their natural enemies (classical biological control) was accepted by many as a logical pest control tactic, but the inverse of this, the possibility that the disassociation (e.g., as by broad-spectrum insecticides) of natural enemies from their native hosts might permit the explosion of the latter to severe pest status, seems not to have occurred to very many persons.

One is forced to this conclusion by the events that took place at the time of the DDT breakthrough and in the years thereafter. Entomological history clearly reveals that at the outset of the synthetic organic insecticide era (i.e., the late 1040s) the great majority of those involved in insect control plunged ahead with the new chemical tools essentially oblivious to the ecological (and genetic) pitfalls that lay ahead. There simply was no general realization that the materials had inherent faults that predetermined the development of serious problems. Even the early signs of impending trouble—unprecedented outbreaks of spider mites, scale insects, and various lepidopterous larvae following the use of DDT, and the rapid development of resistance to that material in the housefly and certain mosquito species—failed to create widespread recognition of what was impending. These developments were simply viewed by most entomologists as minor aberrations that could be corrected by the substitution or addition of other chemicals. Now, a quarter of a century later, the problems have become too varied, widespread, and serious to be further ignored. It is quite apparent that the very characteristics that made the new insecticides effective insect-killers have led to the troubles arising from their use (Huffaker, 1971). *Similar scenarios are developing in weed control, plant pathology, and nematology.*

The basic flaw of modern insecticides is their broad toxicity. The materials kill indiscriminately, and when they are applied in the field, their broadly toxic action can virtually strip the treated areas of arthropod life. Thus, where they are used, the ecosystem web is often shattered overnight, creating a biotic vacuum in which unstable population reactions are almost inevitable. The overall problems would not have been so serious if the development and exploitation of the synthetic organic insecticides had been broadly steeped in ecological theory, but this was not the case. The spectacular performance of DDT had a catalytic effect on the insecticide industry so that an outpouring of other organochlorine and of organophosphate materials quickly followed. Thus, in a span of only 5 or 6 years following World War II, a wide spectrum of synthetic organic materials of broad-spectrum toxicity was made available. The enthusiasm of researchers and users coupled with the highly effective merchandising apparatus of the agrochemical industry gave great impetus to the widespread field application of these materials (van den Bosch, 1978).

Ecological and Economic Backlash

As some had foreseen, inevitable backlashes quickly occurred everywhere that the chemicals were used. The agriculturalist placed himself on an insecticidal treadmill because of three ecologically based phenomena that resulted from use of the new chemicals: (1) *target pest resurgence,* (2) *secondary pest outbreaks,* and (3) *pesticide resistance.* All are directly related to the disruption of natural control and lead increasingly to a "pesticide addiction" from which it is difficult to withdraw (Stern *et al.*, 1959; Smith and van den Bosch, 1967; van den Bosch, 1978). These concepts are diagramed in Figures 10.1 to 10.5. Thus, target pest resurgence occurs because the insecticide kills not only a high percentage of the pest population but also a large proportion of its natural enemies, permitting the rapid invasion and eventual population explosion of the pest (Figures 10.1 and 10.2). Secondary pest outbreaks occur where insecticides applied to control noxious insects destroy the natural enemies of innocuous species occupying the same habitat; the latter, freed of their biological controls, then erupt to damaging levels

O PEST
X PREDATOR-PARASITE

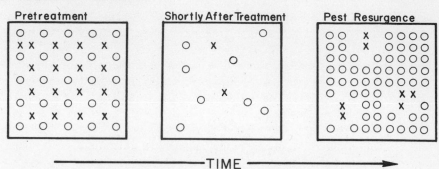

Pretreatment | Shortly After Treatment | Pest Resurgence

───────────────TIME────────────────▶

Figure 10.1. Target pest resurgence. Diagrammatic sketch of the influence of chemical treatment on natural enemy and pest abundance and dispersion with resulting pest resurgence. The squares represent a field or orchard before, immediately after, and sometime after chemical treatment for a pest species. The immediate effect is a strong reduction of the pest, but an even greater destruction of its natural enemies. The resulting unfavorable ratio of pest to natural enemies permits a rapid resurgence of the former to damaging abundance (after Smith and van den Bosch, 1967).

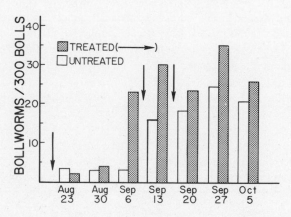

Figure 10.2. Target pest resurgence following applications of a "control" insecticide. In this experiment, plots treated with an organophosphate insecticide federally registered for bollworm control, suffered heavier infestations than untreated plots. Simultaneous samplings of predators revealed that the insecticide destroyed bollworm predators, which permitted resurgence of the pest. The data are taken from an experiment conducted at Dos Palos, California, in 1965.

(Figures 10.3 and 10.4). The survival rate of the pest after the pesticide activity dissipates is quickly changed from a very low rate to a high one. Figure 10.4 illustrates the secondary pest outbreak of the beet armyworm, *Spodoptera exigua,* when insecticide applications were made for the target pest, *Lygus hesperus.* Target pest resurgence and secondary pest outbreaks in

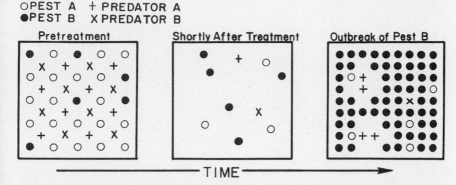

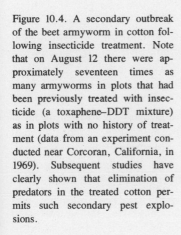

Figure 10.3. Secondary pest outbreak. Diagrammatic sketch of the influence of a chemical treatment on natural-enemy abundance and dispersion with resulting secondary pest outbreak. The squares represent a field or orchard before, immediately after, and sometime after treatment with an insecticide for control of pest A (o). The chemical effectively reduced pest A as well as the natural enemies (+ and ×) but with little or no effect on pest B (•). Subsequently, because of its release from predation, pest B flares to damaging abundance (after Smith and van den Bosch, 1967).

Figure 10.4. A secondary outbreak of the beet armyworm in cotton following insecticide treatment. Note that on August 12 there were approximately seventeen times as many armyworms in plots that had been previously treated with insecticide (a toxaphene–DDT mixture) as in plots with no history of treatment (data from an experiment conducted near Corcoran, California, in 1969). Subsequent studies have clearly shown that elimination of predators in the treated cotton permits such secondary pest explosions.

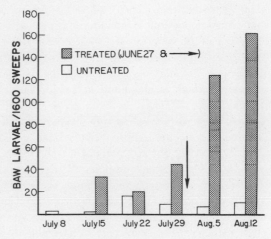

turn contribute directly to and accelerate insecticide resistance because they necessitate heavier and more frequent treatments that speed genetic selection for resistance (Figure 10.5) and often lead to disastrous economic results for farmers (Georghiou and Taylor, 1977; Gutierrez *et al.*, 1979). In addition, the pesticides contribute to pollution and contamination of our air, water, and food.

The seriousness of the trifaceted insecticide treadmill (resurgence, secondary pest outbreaks, and resistance) and its consequence, "pesticide addiction," cannot be exaggerated. It is global in extent and has contributed to

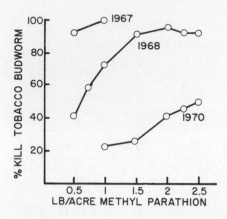

Figure 10.5. Insecticidal resistance. Decline in toxicity of methyl parathion to tobacco budworms from central Texas during the period 1967–1970 (data from P. L. Adkisson, Texas A&M University, College Station).

massive economic and ecological problems virtually everywhere that modern insecticides have been used (Huffaker, 1971). Nothing illustrates this more clearly than the universal intensification of spider mite problems. A quarter of a century ago, spider mites (Tetranychidae) as a group were minor pests. Today they are the most serious arthropod pests affecting agriculture worldwide (Huffaker *et al.*, 1970; McMurtry *et al.*, 1970). For example, in California alone spider mites annually cost agriculture about $60 million, or about one-fifth of all the costs and losses attributable to pest arthropods (Hawthorn, 1969). A single species, the citrus red mite, *Panonychus citri* (McGregor), a minor pest 25 years ago, now costs the citrus industry more than $10 million annually, which is about one-half the total cost of pest arthropods in citrus (Luck *et al.*, 1977).

A wide variety of arthropods prey heavily on spider mites, with predatory mites of the family Phytoseiidae perhaps being the most important group, although tiny lady beetles (*Stethorus* sp.) are also acclaimed by many. Species in several other families of the Acarina also prey on tetranychids. In the Insecta, important spider mite enemies occur in the Coleoptera (Coccinellidae, Staphylinidae), the Neuroptera (Chrysopidae, Hemirobiidae, Coniopterygidae), Hemiptera (Anthocoridae, Miridae, Nabidae, Lygaeidae), and Thysanoptera (*Scolothrips* sp.) (Clausen,1940; Huffaker *et al.*, 1970; McMurtry *et al.*, 1970). Individually and collectively, these predators often effect substantial to complete control of spider mites, and it has largely been the disruption of this naturally occurring biological control that has triggered the global problems.

The worldwide spider mite problem and its causes were thoroughly analyzed (Huffaker *et al* 1970; McMurtry *et al.*, 1970) and presented overwhelming evidence that the globally aggravated spider mite problem is largely a result of disruption by insecticides of the biological controls affecting these pests. In other words, it is the familiar triad of target pest resurgence, secondary pest outbreak, and pesticide resistance. Some spider mites are now so resistant to control chemicals that no available material gives adequate relief. Several researchers worldwide have attempted to develop resistant predator strains and are releasing them in the field with some success (Croft, 1977).

The arthropod pest problems plaguing cotton the world over illustrate equally well the vast importance of naturally occurring biological control. Wherever cotton is grown, it is affected by a variety of arthropod pests, and since it is a cash crop or foreign exchange earner, the first inclination is to protect this source of income from pest depredation. As a consequence, cotton growers are perhaps the world's greatest users of insecticides, and they have paid a severe price for this heavy pesticide use. From virtually every area on earth where cotton is grown, reports have come in recent years of economic, ecological, and even sociological and political problems directly attributed to insecticides (Adkisson, 1971; Smith and Reynolds, 1966, 1972; van den Bosch, 1971). Again as with spider mites, the trouble largely results from the disruption of naturally occurring biological control.

In northeastern Mexico, the tobacco budworm, *Heliothis virescens*, unleashed from its natural enemies by insecticides applied for control of early season pests (e.g., boll weevil, plant bugs), rapidly developed resistance to all insecticides and within a decade totally destroyed a $50 million cotton industry. When the cotton industry was moved to a new area in Mexico, the tobacco budworm destroyed it there in only 4 years (Adkisson, 1971). In California's Imperial Valley, insecticides applied for control of the pink bollworm destroyed the parasites of the cotton leaf perforator, *Bucculatrix thurberiella* Busck, and this pest, which is resistant to virtually all insecticides used in cotton, then erupted to enormous abundance and defoliated thousands of acres (Reynolds, 1971). Resistant tobacco budworms have now also entered the Imperial Valley and during 1978 caused enormous damage. Other areas where particularly severe insecticide associated problems have developed in cotton include the Canete and other valleys of Peru, Colombia, Central America, the Rio Grande Valley of Texas, and cotton-growing areas in Egypt and Turkey (Smith and Reynolds, 1972).

In each of these places, target pest resurgence, secondary pest outbreaks and insecticide resistance have characterized the problem, and again these have resulted in a "pesticide addiction" and were largely derived from the interference of the insecticides with naturally occurring biological control. Some of the important predators involved in cotton are shown in Figure 10.6.

The spider mite and cotton pest problems only partially depict the worldwide biotic backlash to modern insecticide usage. Similar developments have occurred in many other pest control situations (Huffaker, 1971, 1980). However, space does not permit a broader treatment of the matter here. Instead, it is hoped that the preceding discussion has provided insight into the

Figure 10.6A. Crab spider.

Figure 10.6B. Lynx spider, *Lycosa* sp.

Figure 10.6. Natural enemies of cotton pests that are often destroyed by pesticide applications, resulting in resurgence and/or secondary pest outbreaks (F. E. Skinner).

Figure 10.6C. Damselbug, *Nabis* sp., feeding on larva of *Spodoptera exigua*.

Figure 10.6D. Big-eyed bug, *Geocoris* sp.

Figure 10.6E. Ground beetle, *Calosoma* sp.

Figure 10.6F. Minute pirate bug, *Orius tristicolor* (White).

scope and magnitude of naturally occurring biological control and the critical need to take this great natural force into account in future insect control schemes.

Integrated Control

The expanding problems generated by the adverse effects of the modern insecticides on natural enemies have led not only to an unprecedented appreciation of naturally occurring biological control, but also to an increasing realization that a philosophy of insect control that is dominated by a single—in this case, ecologically disrupting—tactic is programmed for failure. Consequently, a new and more sophisticated insect control philosophy has emerged, and programs based on this philosophy are coming into increasing application. The emergent concept has been termed *integrated control* (Stern *et al.,* 1959; Smith and van den Bosch, 1967), while its more recent counterpart is called *integrated pest management* (IPM)(Huffaker, 1980). This latter concept has put greater stress on the need to understand crop dynamics and its influence on pest numbers and vice versa.

Integrated control is a pest population management system that utilizes all suitable techniques (and information) either to reduce pest populations and

Figure 10.G. Larva of the lacewing, *Chrysopa carnea,* preying on a beet armyworm larva.

Figure 10.6H. *Hyposter exigua* (Vier.), a parasite of bollworm, beet armyworm, and other lepidopterous larvae.

maintain them at levels below those causing economic injury, or to so manipulate the populations that they are prevented from causing such injury, by utilizing and blending cultural, biological, and *as a last resort,* chemical controls. Its goal is not simply pest annihilation, but more important, the reduction of pest populations to levels compatible with the economic production of the crop and the concurrent maintenance of environmental integrity. In other words, the ultimate goal of any integrated control (or IPM) program is economical and ecologically acceptable management of pest insect populations.

Integrated control emphasizes the fullest practical utilization of the existing regulating and limiting factors in the ecosystem, and it is this fact that gives it uniqueness. Perhaps the greatest advantage of the approach is that, in maximizing the role of naturally occurring mortality and regulating factors, it automatically assures a higher level of environmental quality, since such maximization can only be attained where there is minimum toxic disruption of the environment. A second major advantage is its economy, which again derives from its heavy reliance on natural controls and minimal dependence on costly artifical measures.

An Economic Perspective

There are two basic considerations in the development of an integrated control program: (1) the degree of ecosystem disruption; and (2) valid *economic thresholds* for the pest species. The economic threshold as defined here is the amount of damage prevented equal to the cost of the insecticide application. Economic thresholds are not static, rather they fluctuate in relation to time, pest numbers, and all other factors impinging on crop production and market price. In a more complete sense, the economic threshold should also include the ecological, sociological, and external economic and aesthetic costs created by any artifical means (e.g., insecticides) used for control (see Gutierrez and Wang, in press).

Establishing economic thresholds is a demanding but necessary scientific endeavor. By contrast, conventional insect control centers on maximum destruction of the pest species, and—except for consideration of the direct hazard to man, domesticated animals, some wildlife (particularly sport fish and game), and perhaps the honeybee—other important elements in the ecosystem are ignored. Economic justification for use of insecticides is invariably poorly established or not established at all. Only too frequently, the essential criterion for pesticide use has been the highly simplistic one of mere pest presence, and the hallmark of pesticide efficiency is maximum kill.

However, society now recognizes that social and environmental costs must be included in any responsible program, even though their inclusion makes the analysis much more difficult and the solution more difficult to implement. There are many points of view on the nature of social and environmental costs, with each additional point of view adding to the complexity of the problem. Crop managers usually seek to maximize profit (but often wrongly use maximum yield as a surrogate), and this notion can be expressed as follows:

$$\Pi_{max} = Profit_{max} = B(X_1, X_2, \ldots X_n)$$
$$- C(X_1, X_2, \ldots X_n) - S(X_1, X_2, \ldots X_n)$$

where B is total revenue given the use of various X_i management tactics, C is the total cost of implementing the X_i management tactics, and S is the social costs of the technologies (e.g., pollution).

The solution of the above profit equation falls on an $n+1$-dimensional surface (n being the number of management tactics), wherein some combination of Xs yields maximum profit. The problem can also be formulated as one of minimization of any of the X_i decision variables.

Figure 10.7 depicts the larger economic problem, including social costs, where the total revenue is the circle. Currently, farmers merely pay for

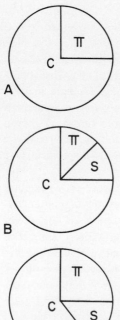

Figure 10.7. Profit (Π) is depicted as the difference between revenues (the area of the circle) and the various production (C) and societal costs (S = negative impact of production practices on society at large). (A) Societal costs are ignored by the farmer. (B) Societal costs are paid by the farmer. (C) IPM practices are substituted for production costs and the farmer pays for societal costs. Note that profit in A equals profit in C.

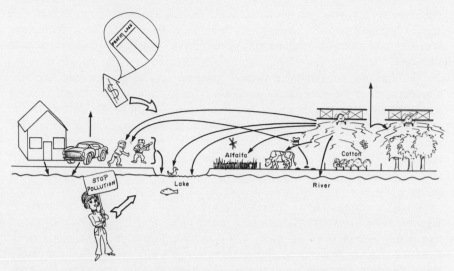

Figure 10.8. Some of the conflicts between private goals (e.g., profit maximization) and societal goals (e.g., enhanced environmental quality). The arrows represent sources of pollution.

production costs (C) (Figure 10.7A) while society shoulders the burdens for health and environmental costs (S). If farmers paid for these social costs, the profit margin (Π), assuming no compensating increase in market price, would be adversely reduced (Figure 10.7B). The best solution to this problem is to develop integrated control programs that maintain current profits without increasing costs to consumers (price) and to substitute good information for unnecessary agronomic inputs (e.g., pesticides, fertilizers) that can result in social losses (Figure 10.7C).

In practice there are many other conflicting points of view between the different parties concerned with agroecosystems. Farmers wish to maximize profits and the public wants abundant, inexpensive food and a clean environment. At another level of complexity, the arrows in Figure 10.8 depict some of the sources of pollution and the nontarget systems affected. It is obvious from the illustration that the public also pollutes, and in addition it competes for many resources required by farmers to produce their crops(e.g. water, land).

The management decision is concerned not simply with the quantity and timing of inputs but also with the consequences of management practices from both private and public points of view. This requires more sophisticated analyses of the crop ecosystem, other interacting crop ecosystems, and the environmental and social constraints surrounding the ecosystem. Each of the variables in the profit maximization equation above [e.g., revenue (B), input costs (C), and social costs (S)] may itself be a very complicated function and may take several years of intensive field research and analysis to define. For each subcomponent, all other factors must be considered. The mathematical description of those interactions is a *model* of the system; an agroecosystem in this case.

In initiating an integrated control program, the automatic first step is a consideration of the ecosystem, particularly as regards the possible effects of any artificial control practices on it, and an assessment of the factors in the ecosystem that favor or mitigate against populations of pests or potential pest species. The second step is the critical analysis of the "pest" status of the reputedly injurious species, to determine which are real as opposed to induced pests. The integrated control philosophy rejects the use of insecticides simply because pest insects occur in an area or threaten to occur there. Instead, under integrated control, criteria are developed to aid pest control advisors to pinpoint those times and places where insecticides are truly needed. It follows that where such need exists, every effort is made to use the most economical, most safe, and most ecologically tenable materials effectively. To help clarify the differences between conventional insect control and integrated control, Table 10.1 compares the two types of programs for cotton insect control in California's San Joaquin Valley.

Research in Integrated Control

The development of the integrated control program in San Joaquin Valley cotton has embraced a wide range of studies in a highly coordinated program: analysis of cotton growth characteristics and fruiting patterns relative to insect phenologies, feeding habits, and population levels; investigations of insect host affinities; analyses of distribution, migration, and phenology of *Lygus hesperus*; studies on the phenologies and in-field and on-plant distribution of *Heliothis zea* and the other lepidopterous species; analyses of the natural-enemy complexes affecting the various pest species and the impact of these enemies on their hosts; studies on nutritional augmentation of natural enemies; determination of the effects of various chemical insecticides on the natural enemies and the cotton plant itself; testing of chemical insecticides and microbial agents on the several pest species; investigations on timing of the various chemical and microbial control treatments; studies on the effect of the feeding of the various pests on cotton growth and fruiting characteristics; analysis of the effects of irrigation and fertilization on populations of *L. hesperus*; and studies on effects of any of these factors and others that might affect yields and profits.

Data in such a program are best obtained in a planned, coordinated way. This requires close cooperation of workers in many disciplines (e.g. agronomy, entomology, physiology, mathematics, economics, computer science).

TABLE 10.1. Comparison of Conventional and Integrated Pest Control Programs in Cotton in California's San Joaquin Valley[a]

Pest	Conventional control	Integrated control
Lygus hesperus	Treatment threshold: 10 bugs/50 net sweeps at any time of season from about June 1 until about mid-August, or prophylactic applications based on time of season, grower apprehension, sales advice, etc.	Treatment threshold: When population levels of 10 bugs/50 net sweeps are recorded on two consecutive sampling dates (3- to 5-day interval) during flower budding period (approximately June 1 to early July). Treatments after mid-July are discouraged as being essentially useless and highly disruptive to natural enemies of lepidopterous pests.
	Cultural controls: None	Cultural controls: In limited use. 1. Strip-harvesting of alfalfa hay fields to prevent migration of bugs into adjacent cotton when alfalfa is mowed. 2. Interplanting of alfalfa strips in cotton fields to attract Lygus out of the cotton. Both practices are based on the strong preference of L. hesperus for alfalfa over cotton.
Lepidopterous pests[b]		
Heliothis zea	Treatment threshold: 1. 4 small (1/4" or less in length) larvae/100 inspected plants. This is a long-standing treatment level of unknown origin that has been found to be completely erroneous. Many fields in the San Joaquin Valley sustain infestations of this magnitude and consequently there is much unnecessary treatment for bollworm control. 2. Nebulous criteria such as time of season, grower apprehension, sales pressure, etc. are also involved in insecticide treatment decisions for bollworm control.	Treatment threshold: 1. 15 small larvae/100 inspected plants where field has been previously treated with insecticides. 2. 20 small larvae/100 inspected plants, where fields have not been previously treated with insecticides. Infestations of this magnitude rarely develop in untreated cotton fields in the San Joaquin Valley. Consequently, cotton that has reached the period of major H. zea activity (i.e., late July and early August) free of insecticide treatment is virtually unthreatened by this pest.

(continued)

TABLE 10.1. (continued)

Pest	Conventional control	Integrated control
Trichopulsia ni	*Treatment threshold:* There is no established economic threshold for this "pest." Instead chemical control is invoked on the basis of such nebulous criteria as "when abundant," "when damage is evident," "when defoliation is threatened." The cabbage looper is commonly abundant in conventionally managed cotton fields where it erupts in the wake of early and midseason insecticide treatments, particularly those applied for *Lygus* control. Consequently, San Joaquin Valley cotton is heavily treated for its control. The insect is very difficult to kill and, as a result, heavy dosages of insecticide mixtures are used against it. Needless to say, such treatments add substantially to pest control costs and they increase the health hazard and environmental pollution.	*Treatment threshold:* There is no evidence that this insect causes economically significant damage to cotton. It is rarely abundant in integrated control fields, since its populations are normally controlled by predators and parasites. Consequently, in fields that have been in the integrated control program, there have been no insecticidal treatments for cabbage looper control.

Spodoptera exigua	*Treatment threshold:* There is no established treatment threshold for this species, and the same criteria are utilized in making decisions for its chemical control as with cabbage looper. However, ongoing research indicates that where abundant, the insect can cause economically significant injury to cotton, but no specific treatment threshold has yet been developed. The beet armyworm is essentially a secondary outbreak pest, and like cabbage looper it is very difficult to "control." Consequently, in conventionally managed fields it often receives costly, pollutive chemical treatments.	*Treatment threshold:* The beet armyworm has not been abundant in the integrated control fields, and there has been no need to consider its control.
Spider mites (Tetranychidae)	*Treatment threshold:* There are no established economic thresholds for spider mites in San Joaquin Valley cotton. A number of selective and nonselective acaricides are used prophylactically and therapeutically for spider mite "control." Again, as in cabbage looper and beet armyworm control, the treatment criteria are largely nebulous.	*Treatment threshold:* No established economic thresholds. Research on this matter is currently under way, as is research on selective chemical controls and the development of spider mite-resistant cotton. In the fields that have been under the integrated control program the growers have at times used a selective acaricide prophylactically.

[a]Lygus bug is the *key pest* in San Joaquin Valley cotton. Intensive computer-aided analyses of economic threshold and factors related to this problem are currently under way.
[b]The complex of lepidopterous pests in San Joaquin Valley cotton includes three major species: *Heliothis zea* (bollworm), *Trichoplusia ni* (cabbage looper), and *Spodoptera exigua* (beet armyworm). All three species are normally under heavy pressure from natural enemies and usually only erupt to great abundance in the wake of insecticide treatments.

Furthermore, as data are gathered, they are analyzed and stored in a central computer file for use by the entire research group. The tools of mathematics, engineering, and systems science, as well as biology, are being brought to bear on complex problems in several commodities worldwide with the goal of developing optimal pest control and production methodologies. Currently, models for cotton and alfalfa and many of their pests are being field tested in integrated control projects in the San Joaquin Valley of California. The project seeks to study not only arthropod pests but also pathogens, weeds, nematodes, and other pests, as well as the crop plants and their interactions as modified by weather.

The integrated control program for cotton and alfalfa in the San Joaquin Valley is not yet a finished product, but where it has been brought into its fullest application, insecticide usage has been reduced by more than 50%. This of course has brought economic benefit to the growers and has correspondingly reduced the pollution threat from insecticides. Unfortunately, herbicide use is increasing worldwide.

Throughout the world, integrated control programs are being developed for human health (mosquitoes), numerous crops (alfalfa, cotton, apples, citrus, peaches, pears, grapes, oil palms, rubber, and cocoa), and animal husbandry (sheep, cattle, and poultry) (Huffaker, 1971, 1980). In each of these cases, preservation and/or augmentation of natural enemies has been a critical factor in the program. Currently, many additional programs are under development, and these too place major emphasis on the role of naturally occurring biological control.

The pattern of today's pest control research clearly reflects the expanding emphasis on integrated control (integrated pest management) and the associated recognition of the importance of naturally occurring biological control. In this light, it is quite apparent that as time passes the research effort focused on naturally occuring biological control and its utilization, not only in insect pest control, but in the control of other classes of pests, will almost surely equal or perhaps even overshadow the efforts in classical biological control. But whatever the case, it is highly unlikely that naturally occurring biological control will ever again be generally ignored in insect control considerations, especially in integrated control.

References

Adkisson, P. L. 1971. Objective uses of insecticides in agriculture. In: J. E. Swift (ed.) Agricultural chemicals—harmony or discord for food–people–environment. Univ. Calif. Div. Agric. Sci. pp. 43–51.

Clausen, C. P. 1940. *Entomophagous Insects*. McGraw-Hill: New York. 688 pp.

Croft, B. A. 1977. Resistance in arthropod predators and parasites. In: D. L. W. Watson and A.

W. A. Brown (eds.) *Pesticide Management and Insecticide Resistance.* Academic Press: New York. pp. 337–393.

Georghiou, G. P. 1971. Resistance of insects and mites to insecticides and acaricides and the future of pesticide chemicals. In: J. E. Swift (ed.) Agricultural Chemicals—harmony or discord for food–people–environment. Univ. Calif. Div. Agric. Sci. pp. 112–117.

Georghiou, G. P., and C. E. Taylor. 1977. Genetic and biological influence in the evolution of insecticide resistance. *J. Econ. Entomol.* **70**:319–323.

Gutierrez, A. P., U. Regev, and H. Shalit. 1971. An economic optimization model of pesticide resistance: Alfalfa and Egyptian alfalfa weevil—an example. *Environ. Entomol.* **8**:101–107.

Gutierrez, A. P., and Y. H. Wang. 1981. Models for managing the impact of pest populations in agricultural crops. In: C. B. Huffaker and R. L. Rabb (eds.) *Ecological Entomology.* Wiley: New York. (In press.)

Hawthorne, R. M. 1969. Estimated damage and crop loss caused by insect/mite pests—1968. Calif. Dept. Agric. E-82-11. 11 pp.

Huffaker, C. B. (ed.) 1971. *Biological Control.* Plenum Press: New York. 511 pp.

Huffaker, C. B. 1980. *New Technology of Pest Control.* Wiley-Interscience: New York. 500 pp.

Huffaker, C. B., M. van de Vrie, and J. A. McMurtry. 1970. Ecology of tetranychid mites and their natural enemies: A review. II. Tetranychid populations and their possible control by predators. *Hilgardia* **40**:391–458.

Luck, R. F., R. van den Bosch, and R. Garcia. 1977. Chemical insect control, a troubled pest management strategy. *BioScience* **27**:606–611.

McMurtry, J. A., C. B. Huffaker, and M. van de Vrie. 1970. Ecology of tetranychid mites and their natural enemies: A review. I. Tetranychid enemies. Their biological characters and the impact of spray practices. *Hilgardia* **40**:331–390.

Reynolds, H. T. 1971. A world review of the problems of insect population upsets and resurgences caused by pesticide chemicals. In: J. E. Swift (ed.) Agricultural chemicals—harmony or discord for food–people–environment. Univ. Calif. Div. Agric. Sci. pp 108–112.

Smith, R. F., and H. T. Reynolds. 1966. Principles, definitions and scope of integrated pest control. *Proc. FAO Symp. on Integrated Pest Control* **I**:11–17.

Smith, R. F., and H. T. Reynolds. 1972. Effects of manipulation of cotton agroecosystems on insect populations. In: M. T. Farvar and J. P. Milton (eds.) *The Careless Technology: Ecology and International Development.* Natural History Press: New York. 373–406 pp.

Smith, R. F., and R. van den Bosch. 1967. Integrated control. In: W. W. Kilgore and R. L. Doutt (eds.) *Pest control: Biological, Physical, and Selected Chemical Methods.* Academic Press: New York. pp. 295–340.

Stern, V. M., R. F. Smith, R. van den Bosch, and K. S. Hagen. 1959. The integration of chemical and biological control of the spotted alfalfa aphid. Part 1. The integrated control concept. *Hilgardia* **29**:81–101.

van den Bosch, R. 1971. The melancholy addition of ol' king cotton. *Nat. His.* **80**(10): 86–91.

van den Bosch, R. 1978. *The Pesticide Conspiracy.* Doubleday: New York. 228 pp.

11

Other Kinds of Pests and Other Biological Methods of Pest Control

Other Kinds of Pests

Biological Control of Vertebrates

On occasion, vertebrate species become pests when they are accidentally or purposely introduced from one area of the world to another. If the new habitat provides no natural population control, the population of the introduced vertebrate expands dramatically and an artificial means of population control must be implemented, either through harvesting or biological control. Examples of such introductions are: the marine toad, *Bufo marinus* L., in Australia, the African clawed frog, *Xenopus laevus* in Florida, the Norway rat, *Rattus* sp., throughout the world, and the European rabbit, *Oryctolagus cuniculus* L., in various regions of the world (Figure 11.1).

The European Rabbit in Australia. The case of the European rabbit in Australia presents a classic example of biological control of a vertebrate pest. It was introduced into Australia in 1859, along with other mammals from Great Britain, in order to make the foreign land more familiar to the new colonists. The native ecosystem had no indigenous predators efficient enough to restrain the rabbit population growth, and by 1900 the population had spread 1000 miles to the north and west. It is thought that the rabbit was transported to many of these regions by trappers wishing to expand their fur business. By the 1930s, the rabbit population far exceeded the demand of the trappers, and Australia experienced a series of rabbit "plagues" (Figure 11.2). As the climatic patterns favored rabbit survival and reproduction, massive population explosions occurred, causing a serious economic problem for the

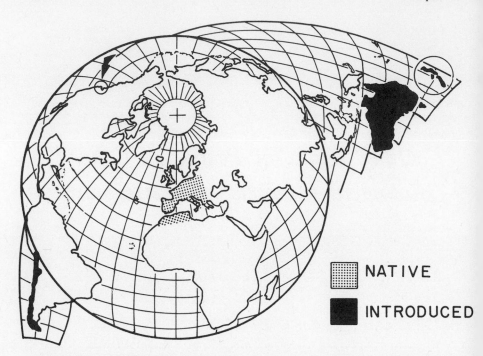

Figure 11.1. The worldwide distribution of the European rabbit, showing its original range and areas where it has been introduced (adapted from Thompson and Worden, 1956).

Australian livestock industry. The millions of rabbits were aggressive competitors for the grasslands, severely damaging the range and farmlands that provided a major source of income to Australians. Trapping, fumigating, and poisoning of the millions of rabbits proved too expensive and yielded little positive effect. The rabbit plagues recurred periodically until the introduction of the Brazilian myxomatosis virus in 1950.

The myxomatosis virus was efficiently vectored by mosquitoes during the spring and summer months, when rabbit reproduction rates are highest, and quickly became a dramatic biological control success, reducing the rabbit population to about 1% of its maximum size. But by 1959, myxomatosis strains varying from high to low virulence had arisen. This evolution reduced its effectiveness in controlling rabbit populations by about 30%. Fortunately, by this time, the introduced European fox and feral cat populations were large enough to exert substantial predation pressure on the less vigorous rabbit population, which counterbalanced the declining effects of myxomatosis. Today, the rabbit population remains controlled by cat, fox, and virus mortality (Davis *et al.*, 1979).

Biological Control of Dung

Biological control has also been used to counteract certain unfavorable secondary effects of introduced domesticated species. A good example is the biological control of cattle dung in Australia by African dung beetles (Waterhouse, 1974) (Figure 11.3). The English colonists in Australia brought sheep and cattle with them as a food supply. The Australian mammalian fauna is composed mostly of marsupials that produce relatively dry pellet-shaped dung. The Australian dung beetle fauna was adapted to aid in the decomposition of marsupial dung but not at all suited for, or even attracted to, the larger, wetter dung pads of cattle. Before the successful project began, cattle dung pads would dry out on the surface of the soil and sometimes persist for years before being broken down by trampling and weathering. As the cattle increased in numbers, more and more acreage of pasture was removed from production due to the accumulation of dung. Because the dung persisted on the surface, and many nutrients were not recycled into the soil, fertility of the pastures dropped, and the number of cattle supported per acre fell at a rate of up to 20% a year.

Figure 11.2. The European rabbit problem in Australia. (A) Populations at an enclosed waterhole. (B) The destruction of pastures as a result of food consumption and burrowing (courtesy D. F. Waterhouse, CSIRO, Australia).

Figure 11.2. (*continued*)

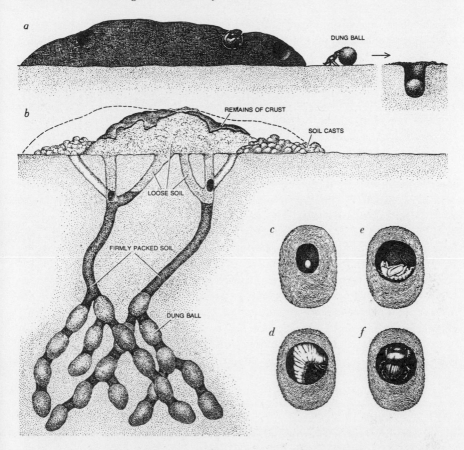

Figure 11.3. Biological control of dung in Australia. Two methods of dung disintegration are shown. (a) One group of dung beetles, represented by such species as *Garreta nitens,* cuts bits of dung out of the pad, forms them into spherical balls, and rolls the balls away to be buried in a shallow pit. (b) Most species instead form their nests in a network of tunnels excavated below or adjacent to the pad. Working together, a male and female dig a tunnel and carry dung down into it. The female forms a ball (ovoid, in the case of *Onthophagus gazella,* shown here), lays an egg in it, and closes the ball (c). As many as 40 balls are formed. The tunnels are backfilled with firmly packed soil; loose soil fills the upper parts of the tunnels and is left on the surface along with some remains of the dung pad. When the larva hatches (d), it feeds on the dung. After passing through the pupal stage (e), the young adult (f) emerges and makes its way to the surface. (Figure and description courtesy of *Scientific American.*) (Waterhouse, 1964.)

In contrast to Australia, Africa supports a wide variety of bovine ante-lopes and other ungulates and has a rich native fauna of dung beetles (Scarabaeidae) adapted to feeding on their dung. In the native areas of South Africa, dung beetles can bury a cattle dung pad within 24 hr. In 1963 several candidate species were brought to Australia from South Africa, and after careful quarantine to eliminate both beetle and bovine diseases, several hundred thousand beetles of four species were reared and released in northern Australia. Within a few years, two species, *Onthophagus gazella* (F.) and *Euoniticellus intermedius* L., became widespread throughout tropical northern Australia and successfully controlled dung during parts of the year. Other species are being brought in to complement these two species during other times of the year, and still others to dispose of the dung in the drier areas of the Australian outback.

Dung control in areas lacking native dung-adapted insects has also been accomplished in New Zealand, Papua New Guinea, Hawaii, and North America. Besides restoring pasturelands and recycling nutrients, dung beetles have also been shown to significantly reduce the numbers of dung-breeding flies such as horn flies, face flies, and bushflies.

The foregoing account illustrates the potential for biological control in not simply controlling a pest but altering the ecology of an ecosystem brought into imbalance by the introduction of exotic organisms (e.g., cattle). There are many other similar problems arising from the introduction of desired exotic species that remain unsolved. A good example is the persistence of leaves of imported eucalyptus trees on the soil in North America and elsewhere. In this case, local detritivores are not adapted to break down and recycle the leaves of the imported trees, resulting in a buildup of dry fuel and often creating a fire hazard during the dry summers.

Other Biological Methods of Pest Control

Besides the use of natural enemies and microbial control, there are several other methods of a biological nature that have been applied with success in the suppression of noxious species. These methods are based on host plant (or animal) resistance to pest insects or diseases, certain types of cultural control, control through use of sterile insects, or other genetic manipulations.

In several cases, one or another of these methods used alone has resulted in the satisfactory suppression of a serious pest. In a few cases, combinations of two or more of the techniques have been used with or without pesticides for control of a given pest. For example, the spotted alfalfa aphid in California was controlled using introduced parasites and resistant plant varieties. In these latter situations, any one method alone usually has not been sufficient to provide fully adequate control.

Host Plant Resistance

The use of crop varieties resistant to attack or damage by pest has been practiced in North America since the later decades of the nineteenth century

(Painter, 1951; Anonymous, 1969). Although the first effective application of this technique did not occur in North America, it still had its roots on this continent. This case involved the grape phylloxera, *Phylloxera vitifoliae* (Fitch), a North American root aphid that was accidentally introduced into Europe. The European grape, *Vitis vinifera* L., was extremely susceptible to the phylloxera, whose attack was so devastating that Eruopean viticulture was threatened with destruction. The use of rootstock of American grape, *V. labrusca* Fox, onto which were grafted European table and wine grape varieties, provided a rapid and effective solution to this problem.

The grape phylloxera-resistant rootstock example is one where man discovered and took practical advantage of the natural occurrence of resistance. However, in many subsequent examples, man not only made use of resistance; he actually created resistant varieties of plants and animals by hybridizing or crossing, and by deliberate, artificial selection. This points us to the basic requirement for the development of resistant varieties, which is that the tendency toward resistance to pest attack or damage must be an inherited genetic trait. The resistance may be due to a single gene (sometimes categorized as vertical resistance, as in the resistance to some plant pathogens) or to a complex of genes (horizontal resistance, as in the resistance to some insect pests with more diverse physiology and behavior). Of course there are exceptions to almost every rule in biology, and there certainly are to this generalization.

Early examples of the development of plant resistance by selective breeding include the creation, by the early 1900s, of potato varieties resistant to potato late blight, *Phytophthora infestans* (Montemartini) DeBary. This disease was responsible for the Irish potato famine of the mid-nineteenth century. By the early 1920s, the selective breeding of wheat varieties led to the creation of new varieties resistant to the major insect pest of wheat in North America, the Hessian fly, *Mayetiola destructor* (Say). Since these beginnings there has been a steady, albeit slow, progress in the development and use of resistant varieties—but as a result of this program, some of our principal, devastating insect and disease pests have been overcome (Painter, 1951; Beck, 1965).

The method used by plant breeders may be generally described as follows. In fields of the crop plant heavily infested with the pest in question, a search is conducted, sometimes over a wide geographical area, for individual plants seemingly free of or little affected by the attack. These plants are removed to the plant breeding nursery or greenhouse, and seeds are produced and collected. Seedlings are grown and crossed with each other to produce hybrid clones of plants. Tests are conducted to assure that resistance is retained, not only to the target pest, but also to other pests to which the previous varieties were resistant. All too often, the latter is overlooked and serious outbreaks occur of such disease as corn blight, which devastated field corn in the midwestern region of the United States during 1971.

A novel method for selecting resistant plants has been proposed by Robinson (1976) and involves selecting and breeding not only economic cultivars showing obvious resistance but also other plants with less desirable economic traits, with the view that components of resistance are also to be found in them. The goal of this approach is to insure consistently high yield by making resistance to pests and adaptedness to edaphic, climatic, and other factors likely to affect plant growth and development as broadly based as possible. The clones are also studied to assure that other necessary plant characteristics are present, such as vigor, yield potential, and growth qualities. Where resistance is inherited and other qualities are retained or restored by additional hybridization with parental stock, the new, selected variety is then provided for general use. Such plant breeding avoids the pitfalls that occurred with the narrowly selected *green revolution* varieties that are high yielding when kept under high inputs of pesticides and fertilizers and good water management but that may produce disastrous yields when one or more of these inputs is deficient. This latter point is extremely important, as these varieties are often released in regions of the world that can ill afford the costs of these imported chemicals and where the modern technology to maximize yields is only marginally developed. Roumasset (1976) provides a provocative discussion of the adverse effects of green revolution rice in Southeast Asia.

Careful study of the nature of insect resistance in crop plants shows that resistance can be ascribed to one or more of three types: (1) *tolerance*, (2) *antibiosis*, or (3) *nonpreference* (Painter, 1951). Tolerance occurs when the resistant plant can sustain normal levels of pest infestation without damage or loss in yield. Antibiosis refers to the suppression of attack or damage through some physiological effect on the pest, such as reduction in size, vigor, or developmental rate of the growing pest, reduction of fecundity of the adult female, or a decrease in survival rate of the pest during its life cycle. Nonpreference refers to the reduction of attack brought about by unattractiveness of the host or inability of the pest to recognize it.

In many closely observed cases of host plant resistance, any two or all three of these types of resistance mechanisms may occur together. Among the wheat varieties resistant to the Hessian fly, the midwestern variety, Pawnee, is tolerant of the pest. With other wheats, antibiosis and tolerance together are involved. An example of nonpreference is the resistance in corn varieties to grasshopper attack, and in certain *Solanum* species to attack by the Colorado potato beetle. The resistance of certain midwestern corn varieties to the European corn borer involves contributions from all three modes of resistance—tolerance and antibiosis in substantial amounts, and nonpreference to a lesser degree.

Some Examples of Use of Plant Resistance

Other successes in the discovery and development of resistance in crop plants involve the wheat-stem sawfly in Canada and the United States; the creation of varieties of corn resistant to fall armyworm, stalk borers, cornleaf aphid, and corn earworm; and wheat, oat, and barley varieties resistant to the cereal leaf beetle and the greenbug. In the case of the corn earworm, resistance in certain corn varieties was of only enough benefit to enable a reduction in the number of pesticide applications needed to control this pest. However, even this outcome can lead to considerable value in reducing the costs of corn production and the side effects of excessive pesticide usage.

One of the more successful cases of host plant resistance development in a large crop devastated by a new insect pest concerns the spotted alfalfa aphid, *Therioaphis trifolii*. This pest invaded North America for the first time about 1954 in the southwestern states of New Mexico, Arizona, and California. Within 2–3 years it had spread throughout 30 other states. The pest thrives on alfalfa, causing the deaths of individual plants, loss of vigor and reduction of growth, and the fouling of cropped hay and harvest equipment by the excessive accumulations of honeydew (aphid excreta).

In several alfalfa-breeding nurseries in Southern California and Nevada, where work was proceeding on the development of varieties resistant to certain plant diseases such as wilt, and to stem nematodes, the spotted alfalfa aphid proliferated. However, among the disease-resistant varieties it was noted that several plants of the variety Lahontan showed resistance to the aphid. These were collected, bred, and crossed to produce a very effective spotted alfalfa-aphid-resistant Lahontan, a variety now used to a considerable extent over much of the southwestern United States. Moapa is another alfalfa variety resistant to the spotted alfalfa aphid and also used in much of the southwestern United States. It was derived, after further hybridizing and selection, from a susceptible variety known as African.

The use of these two resistant varieties of alfalfa, coupled with the biological control of the pest aphid by imported parasites and native predators (Chapter 9), has made possible the continued production of this important forage crop. In Kansas, however, neither Lahontan or Moapa is well adapted to the local conditions. The variety most extensively grown there is called Buffalo. The successful search for and crossing of individual plants of the Buffalo variety found resistant to the spotted alfalfa aphid eventually led to the creation of a new, resistant variety called Cody. This was found to be as good as or even better than Lahontan when grown in Kansas conditions (Painter, 1960).

In most situations, only those growers who adopt resistant varieties of crop plants benefit from the practice, while those who retain the susceptible varieties continue to suffer damage (Painter, 1960). However, in some cases, the use of resistant varieties produced "spillover" benefits. Such is the case with wheat varieties resistant to the Hessian fly in Kansas. With the widespread adoption of resistant wheats, the overall incidence of Hessian fly declined by as much as 50%, leading to reduction of damage even in fields planted to nonresistant wheat.

Host plant resistance is by no means permanent, since the pest itself has the ability to adapt to the new varieties. However, the breakdown of resistance to insects (especially for polygenically derived resistance) in initially resistant varieties is often a slow process, and with appropriate effort it can be counteracted by continuing research for additional resistant stocks. It is, on the other hand, relatively inexpensive once the proper teamwork between plant breeder, agronomist, and pest expert has been developed. Host plant resistance is sometimes the only practical insect control method where biological control agents are not known or effective, and crop production economics do not justify use of expensive pesticides. Unfortunately, all too often each discipline ignores other disciplines, resulting in very narrow suboptimal selections.

But plant varieties resistant to diseases are also numerous. In fact, maize varieties resistant to leaf rust have been known since prehistoric times. Flax varieties were developed from 1908 to 1935 that are resistant to *Fusarium* wilt. Sugarcane varieties resistant to mosaic virus have been in use since 1926. Wheats resistant to wheat-stem rust, *Puccinia graminis* Persoon, although ephemeral in durability, have been in existence, with millions of bushels saved since 1938.

Nematode resistance has also been successfully developed in crop plants including alfalfa, bean, barley, tobacco, potato, rice, and tomato.

Cultural Control

Cultural control means the development or adjustment of agronomic or horticultural procedures to reduce pest abundance or minimize or prevent damage resulting from pests (Anon., 1969). Stated another way, it is the alteration of the agricultural environment so that, while still producing a suitable crop, the environment is rendered less favorable for the pest species. Cultural controls were very likely the earliest control measures developed and applied by man. They include *sanitation* or cleanup of crop residues that may serve as sources of infestation; changes in planting schedules, either through *crop rotation* or use of *mixed crops* grown in relatively small plots rather than solid stands; changes in *planting* or *harvest time*; *strip-cropping* or harvesting of only parts of a field at one time; use of *tillage* or working of the soil; *clean culture* or elimination of volunteer hosts, weeds, litter, or cover crops; and the use of *trap crops* that attract the pest away from the production crop. There are many variations on these basic practices; some of the more effective ones are described briefly in the following sections. Some workers even include the growing of resistant varieties as a form of cultural control, but we prefer to consider this a distinct method.

Sanitation. Sanitation includes the plowing under of infested plants after harvest, as has been done with cotton for control of the pink bollworm and boll weevil; the destruction of prunings of horticultural crops, citrus, apple, and peach to destroy twig- and branch-inhabiting pests; and the gathering and removal of fruit drops or mummies (stick-tights) that harbor overwintering infestations of, say, navel orangeworm, *Amyelois transitella* (Walker). In forests, the disposal of slash resulting from logging activity reduces infestations of bark beetles.

Crop Rotation. Since ancient times, crops have been rotated, that is, planted in a sequence of species or "rotated" with wild growth for purposes of rejuvenating soil fertility (e.g., through use of legumes), suppressing soil erosion, and reducing pest infestations. Crop rotation, which is practicable mainly with field crops, has been used successfully as a means for suppressing plant-specific, soil-borne plant pathogens such as nematodes and fungi. For example, the alternation of potato with alfalfa serves to reduce infestations of wireworms. Rotation of soybean or oats with corn has been widely used to suppress corn rootworms, *Diabrotica* spp.

Mixed Cropping. In ecological terms, increasing the plant diversity of the agricultural ecosystem often aids in the maintenance of pest populations at moderate to low numerical levels. The principal mechanism here is the perpetuation of natural controls, such as general predators and parasites that act as mortality agents for many plant pests. But just as important is the fact that pests must search for hosts, hence their populations are lower.

Mixed planting in modern American agriculture commonly takes the form of interplanting, or in certain instances, the planting in adjacent fields of high and low risk corps vis-á-vis a common insect pest. For example, in California some growers practice interplanting of alfalfa strips between strips of cotton so as to attract *Lygus hesperus* out of the cotton where it can do serious damage, and into alfalfa where it does little harm (Figure 11.4). Unfortunately, farmers do not use these methods as much as they should because of the minor inconvenience it causes and the ready availability of inexpensive chemical control methods. In many countries of Central America, mixtures of different species of food plants such as corn, cassava, and beans may be grown in such a manner as to intercept as much sunlight as possible to utilize the land efficiently, reduce pest and pathogen damage, and ensure that crops will always be available for food. The aim in this case is to stabilize production, not necessarily maximize it. It has been used most extensively in undeveloped areas of the world (e.g. China, Southeast Asia) and by backyard gardeners.

Timing of Planting or Harvest. Early planting of corn in northern regions reduces chances for infestation of *Heliothis* spp. and fall armyworm, which can only overwinter in the southern United States and thus must move northwards each summer. Early corn planting in the southern states reduces infestations of the southwestern cornstalk borer. Late planting of winter wheat each fall markedly reduces oviposition by Hessian fly.

Early harvest and subsequent plant destruction in cotton fields for control of pink bollworm and cotton boll weevil has been mentioned previously

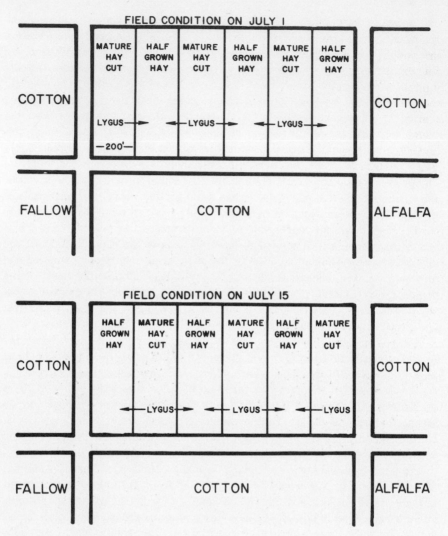

Figure 11.4. A diagrammatic representation of mixed cropping of cotton and alfalfa, as well as the effect of strip-harvesting on the movement of the pest *Lygus hesperus* at two points in time during the growing season.

above. Early-ripening varieties of walnuts are much less susceptible to codling moth in California. Early cutting of the first two crops of alfalfa in northern California destroys a substantial portion of the alfalfa weevil, *Hypera postica*, infestations while the pest is still in the larval and pupal stages and cannot tolerate the loss of food or the physical shock of excessive dryness and heat resulting from the cutting activity.

Strip Harvesting. This control technique is of special value to crops such as alfalfa, which provide several cuttings of hay in a season because strip-harvesting, as contrasted to solid cutting of entire fields, allows the perpetuation of natural-enemy populations and consequently the maintenance of pest populations at relatively low levels Stern *et al.* (see Figure 11.5). However, when alfalfa fields are cut in their entirety at one time, more economical use is made of harvest crews and equipment, and irrigation is facilitated. But under the "solid cut" practice, there is an enormous destruction of the resident insect fauna, including the complex of natural enemies. And as is so often the case, as regrowth occurs, the first insects to reoccupy the field in numbers are the plant pests. Only later do the predatory and parasitic species return and increase in numbers—often too late. By cropping the fields in alternate strips, the insect fauna is maintained in the field and the new alfalfa growth is thus simultaneously populated by both pests and natural enemies. Pests that have been maintained under good control by this practice are the pea aphid, *Acrythosiphon pisum,* the spotted alfalfa aphid, and several lepidopterous species, including the alfalfa caterpillar, *Colias eurytheme,* the western yellow-striped armyworm, *Spodoptera praefica* and the beet armyworm, *Spodoptera exigua.* This practice has two additional advantages: (1) alfalfa in the central valley of California is a vast insectary for natural enemies that invade other crops such as cotton, and strip-cropping serves to keep their numbers high; and (2) this practice helps retain pests, e.g., *Lygus hesperus,* that would migrate to the less preferred crop if the entire field were cut.

Tillage. The destruction of all weeds and border plants near a crop field and along farm roads, irrigation ditch banks, and the like is a common practice in many parts of the United States. The benefits of such a practice are debatable.

On one hand, such "scorched-earth" practice eliminates many sources of pest infestations or pest overwintering sites, and has been claimed to reduce populations of the sorghum midge, *Contarinia sorghicola* (Coquillett); the squash bug, *Anasa tristis* (DeGeer); the harlequin cabbage bug, *Murgantia histrionica* (Hahn); the green peach aphid, *Myzus persicae* (Sulzer); and the Japanese beetle, *Popillia japonica.* The method has been most successful in control of overwintering boll weevil populations in Texas. The elimination of these weeds and "escaped" crop plants prevents the pest from maintaining itself at low levels at the periphery of the field or orchard, from which reservoir the new growth of the agricultural crop is invaded as the growing season advances.

On the other hand, the elimination of all weeds and other plants from the periphery of the agricultural field destroys habitats, alternate hosts, food sources, and overwintering refuges of a great variety of natural enemies and other beneficial insects, such as pollinators. Predators and parasites can be

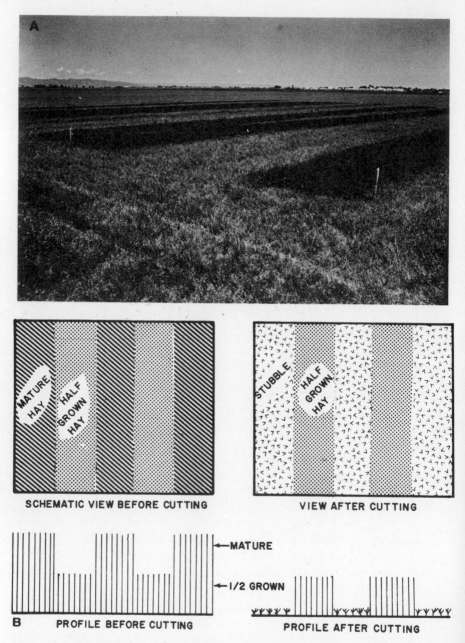

Figure 11.5. Strip-harvesting of alfalfa for the enhancement of beneficial species and control of pests. (A) A photograph of a strip-cut field in the San Joaquin Valley. (B) A diagram illustrating strip-harvesting.

harbored on colonies of their preferred prey or hosts, ready to move with them into the fields each spring. Where these natural enemies have important roles in the biological control of the given pest, the interference of their stability and presence by clean-culture practices leaves the field to the exclusive play of the pest species. The consequences have been well documented: rapid pest population buildup following pest reinvasion, with the natural-enemy populations following much too late to prevent a great deal of early and midseason pest damage.

Because there are both positive and negative values following the practice of clean culture, each pest situation must be worked out on its own merit. Certain phases of current research in biological control attempt to retain and augment natural-enemy populations in agricultural situations, particularly with reference to field crops, by maintenance of overwintering sites, alternative host habitats, and predator "nesting sites"—either by fostering weedy or bramble-type hedgerows on roadsides or by provision of artificial refuges, bird boxes, wasp nests, coccinellid hibernation devices, and the like.

Trap Crops. Trap crops are plantings in or adjacent to the agricultural field on which the pest population congregates. These trap crops are, for this reason, composed of highly preferred host plants. Since the trap crop is primarily of use as a pest control means, it may be the site of heavy pesticide treatment, or destruction by fire or the plow, or any other procedure aimed at pest elimination. The potential of this method for early and late season suppression of the bean leaf beetle [*Cerotoma trifurcata* (Foster)] in Louisiana has been documented; its use for boll weevil control has not been fully exploited. In the latter case, boll weevil emits a pheromone that attracts other weevils to small plots of early fruiting cotton where they can be destroyed with minimal environmental disruption before the large scale adjacent crops begin fruiting.

In Hawaii, trap crops of corn surrounding fields of melons or squash prove highly attractive to adult melon flies. The treatment of trap crop corn rows with pesticides kills many pests yet leaves no chemical residues on the melon or squash crops, nor does it interfere with the natural-enemy fauna present in the field.

As has already been mentioned, the use of alfalfa strips within cotton fields serves to attract and concentrate lygus bugs that would otherwise attack and injure the cotton plants. Unfortunately, this method in alfalfa and other crops is not widely used because its value is not fully recognized.

The Sterile-Insect Control Method (Sterile-Male Technique)

The sterile-insect control method, also called the autocidal or self-destruction technique, or more popularly the "sterile-male" technique, is one of the most ingenious pest control procedures yet developed (Knipling and

Klassen, 1976; Anon. 1969; LaChance *et al.*, 1967). In brief, it involves first artificially sterilizing large numbers of pest individuals (*usually males*) by means of gamma radiation or chemosterilants and then releasing them into an untreated or wild population habitat where the treated adult males mate with wild females. The result is a substantially lowered progeny production for the next generation. By repeating this procedure for a series of consecutive generations, the wild population declines to eventual annihilation. A necessary requisite of this method is an extremely low target population before it can be initiated over a large geographic area. The low population greatly enhances the probability that sterile males will mate with wild or native females.

For the technique to be applied, as it was first successfully done against the screwworm fly, *Cochliomyia hominivorax* (Coquerel), in the southeastern United States in 1958, knowledge must be developed concerning: (1) sterilizing the male adults without simultaneously reducing their competitive reproductive vigor relative to wild male adults; (2) rearing or otherwise accumulating very large numbers of males of the species for treatment and release; (3) accurate censusing of the population of the target insect with respect to both absolute numbers and seasonal numerical fluctuations; (4) estimating the rate of potential population increase from generation to generation in order to arrive at suitable "overflood" ratios of treated to wild males; (5) ascertaining that any potential harm that might result from the release of large numbers of pests, perhaps including some not completely sterilized males, will not outweigh the benefits to be realized by any subsequent annihiliation or long-term suppression of the species;* and (6) determining that the costs of carrying out such a control program relative to the costs of other pest suppression techniques and/or to the value of the protected commodity remain favorable.

The possibility that the sterile-insect technique can literally eradicate an entire pest population from an area and thus provide a permanent solution to the particular pest problem under consideration has attracted a great deal of resources, theoretical attention, and practical evaluation of the several aspects of the technique. The limitations are of two kinds—technical and ecological.

The sterile-insect technique of pest control has been applied successfully against the screwworm fly on the island of Curacao, in Florida, and in other southern states. In the first two localities, the programs led to complete annihilation, but in the southwestern states, a continuing program of pest suppression to subeconomic levels is required. On the small island of Rota in the western Pacific, near Guam, the sterile-insect technique has led to the eradication (at least for a number of years) of two different species of fruit flies: the melon fly, *Dacus curcubitae* Coquillett, and the Oriental fruit fly, *D. dorsalis*. While tropical fruit flies such as these are noteworthy for their very

---

*Irradiated fertile females from a Peruvian source were accidentally released in 1981 in an eradication program against the Mediterranean fruit fly in California.

high abundance in areas normally infested, in these two cases, when eradication was undertaken on Rota, the two species were present at unusually low levels because of scarcity of host fruits, and in the case involving the melon fly, because of prior suppression with a toxic bait.

The sterile-insect procedure has also been used in pest prevention. In this modification, sterilized insects are released in an area not actually infested but yet subject to potential or even occasional invasion by that particular species. Such has been the application against the Mexican fruit fly, *Anastrepha ludens* (Loew), a pest that normally occurs in may places in Mexico but occasionally reaches and sometimes even crosses the international border at California. Continuous releases of sterilized males in a zone along the border serves to overwhelm any movement of the wild fly population northward. The same approach is also being attempted in the southern San Joaquin Valley of California as a preventive measure against spread of the very destructive pink bollworm in cotton. These two programs are of questionable value.

The sterile-insect technique, using either irradiated or chemosterilized individuals, has been applied on certain occasions, mostly on an experimental basis, against small segments of large pest populations with usually very positive results in terms of reduction of numbers. However, such results by themselves can be misleading relative to potential annihilation or economic control. Unless the treated subpopulation is isolated (or isolatable) from other subpopulations, or unless the technical feasibility for annihilation or control of the entire regionwide population has been established, such results do not themselves constitute successful application of the technique.

Drawbacks to the Procedure

Effective sterilization of the adult of a species (males and/or females) must be attained without impairing suitable levels of mating competitiveness in the insect. In many cases, preliminary research to establish such poststerilizing reproductive competitiveness indicates that the method is often unfeasible.

The ability to rear the pest organism by artificial means in very large numbers constitutes another technical barrier. It so happens that in the successful autocidal control programs, the species concerned lent themselves to mass culture. For the screwworm eradication program in Florida, production levels of over 50 million male flies per week were accomplished. For the southwestern United States–Mexico suppression program, 100–150 million flies per week were reared. For eradication of the two *Dacus* species from Rota, sterile male releases amounted to about 10 million per week. In all these cases, the pests were amenable to rearing in mass situations on artificial media, under conditions of excessive crowding.

Hence the sterile-insect technique will very likely be impracticable for pests that cannot be sterilized without causing excessive behavioral alterations, for those species that can only be reared on living plant material, those highly cannibalistic as larvae, or those of excessive susceptibility to disease, which must be reared under aseptic conditions. Individuals that develop high levels of fitness under laboratory conditions are invariably unsuited to face the stresses of nature, and hence may contribute to the failure of the method.

Nevertheless, there are a number of species besides the ones mentioned above that may very well be susceptible to the sterile-insect technique. Given the appropriate ecological circumstances of isolation and wild population accessibility, the Mediterranean fruit fly, *Ceratitus capitata,* the codling moth, *Laspeyresia pomonella,* and similar major pest species may well be appropriate for such a control approach. There has been considerable interest in the eradication of other key pest species such as the boll weevil and pink bollworm, but the evidence to date indicates that those problems would be much more ecologically difficult and costly, and hence not feasible. The boll weevil program must rely not only on sterile male releases but also on the prior use of a pesticide to reduce the population to sufficiently low levels for the procedure to appear feasible. The programs would be extremely costly. For example, it is estimated that complete eradication of the boll weevil, if accomplished, would cost more than a billion dollars, and intensive, widespread insect pesticide use would be required. Is society willing to pay this staggering price, especially if alternative methods such as cultural or biological control, the use of resistant varieties, or general integrated control can accomplish satisfactory results much more cheaply? Already, short-season cottons have been shown to greatly reduce boll weevil losses, decrease pesticide use, and greatly increase profits in Texas, and the same results would accrue for pink bollworm in California, Arizona, and New Mexico.

Genetic Control of Pests

The genetic control of pests involves the alteration of the genetic makeup of pest organisms such that they become less vigorous, less fecund, or genetically sterile as a consequence of hybridization, resulting in their decline in numbers to either low densities or extinction (Anonymous, 1969). The term *genetic control* has been amplified by some to include host plant resistance on the one hand and the sterile-insect release principle on the other. We prefer to restrict the scope of genetic control to some artificial influence by man on the gene composition in the target population in a way that allows the spread of the deleterious genes through the population, rather than cutting off reproduction immediately, as with the sterile-male technique. Host plant resistance and sterile-insect methods have been treated previously in separate sections.

Many basic properties of insects lend themselves to genetic manipulation. Species of insects appear in numerous polytypic forms, biotypes, races, and strains. On an evolutionary time scale, species of insects are genetically altered by selections, either natural or artificial, this genetic plasticity being evidenced by their highly varied and variable adaptations to different ecological situations. On a shorter time scale, their life histories are highly suited for genetic manipulation by scientists, as they have short generation times and relatively high reproductive potentials.

Research on the genetic control of insect pests has centered on (1) hybrid sterility, (2) cytoplasmic incompatibility, (3) conditional lethals, and (4) growth alterations. However none of these has yet led to large-scale field testing, let alone actual pest control.

Hybrid Sterility. Hybrid sterility is the result of the hybridization of two different but related species, with the production of sterile progeny. Where these hybrids are fully viable, they remain in the environment to compete with normal individuals, with a consequent reduction in the breeding population.

This scheme requires the determination of species that will cross, one being the target pest and the other a nonpest. Mass culture and release of the nonpest species in the target pest area will bring the control method into action.

This method has been proposed for the two tsetse fly species, *Glossina swynnertoni* Austen and *G. morsitans* Westw. Where, as in this case, both species are pests, mass culture and hybridization will have to be carried out in "insect factories" with release only of the sterile hybrids. To this extent, the procedure is equivalent to the sterile-insect control method.

Cytoplasmic Incompatibility. Certain strains of the same species when crossed produce either sterile offspring, or fertile and sterile offspring together (resulting from reciprocal crosses), or sterile males. The sterility in any case is the result of the destruction of the sperm as it penetrates the ovum.

This technique has been proposed for a number of mosquito species [e.g., *Culex pipiens pipiens* L. and *Aedes scutellaris* (Walker)] where these strains have been found. One of two modifications can be used; production and release of males of one type into the area of occupancy of an incompatible type, or production and release of the hybrid, which will compete sexually with the target pest.

Conditional Lethals. Certain alleles of genes controlling adaptiveness to temperature, for example, produce normal effects under certain temperature ranges, but at higher or lower temperatures produce nonadaptive or lethal effects. Again, this scheme derives from observations of mosquitoes, in this case the yellow fever mosquito, *Aedes aegypti* (L.). A strain of this species has been found in which, when it is reared at 27–28°C, half the progeny are normal females and the rest intersexes. The latter resemble normal females and will mate and receive sperm, but produce no eggs. Any control procedure using this technique will require the introduction of large numbers of the maladapted strain into the target pest environment.

Whitten (1970) proposed a method for using conditionally lethal genes of the sort previously described. In his procedure, the strain of pest possessing the conditional lethal allele is irradiated to produce chromosome translocations. So long as this translocation strain is maintained in a homozygous state, it remains reproductively fertile. Mass culture and release of such a strain enables it to hybridize with the wild or pest strain to produce the heterozygous state. These hybrids are sterile. By suitable overflooding of the wild population the latter eventually is totally displaced. Then, with a shift in the seasonal temperature, the conditional lethal comes into play and results in the self-destruction of the released strain. The overall result could be the full elimination of both strains from the target environment.

Growth Alterations. In some species of insects, the presence or absence of diapause is known to be under genetic control, as for example in the silkworm, *Bombyx mori* (L.), and the cotton boll weevil, *Anthonomus grandis.* By use of nondiapause genes in a "translocation" strain, a normal, diapausing strain can be displaced as per Whitten (1970), discussed previously. When the winter (or summer) diapause season arrives, the released strain then is annihilated.

Another procedure for genetic manipulation of a diapausing population has been considered in the case of the field cricket, *Teleogryllus commodus* (Walker) in Australia. In the southerly, more temperate regions this species produces diapausing eggs; in the more tropical north it produces only nondiapausing eggs. Northern males crossed with southern females produce nondiapausing eggs. Hence, mass culture and release of northern males into southern infestations should result in the production of numerous eggs unable to carry over the winter. As in the sterile-male technique, repeated releases conceivable could produce substantial and even complete reduction of the southern strain of the cricket.

Summary

All of these methods show great ingenuity, but some are more feasible than others. Among the more feasible methods, in addition to biological control, are host plant resistance and cultural controls, including sanitation. Crop rotation of soybean and corn could for example eliminate corn rootworm as a major pest in much of the 64 million acres of field corn in the United States and greatly lessen pesticide pollution. Unfortunately, economic considerations (e.g., machinery cost, commodity prices) greatly restrict the adoption of this practice since farmers in the corn belt, in particular, find growing corn year after year to be more profitable.

Methods such as strip-cropping and trap cropping are thought to require unwarranted extra investments of time and resources and have been little adopted partly because of this. Under current agricultural practices, mixed cropping in much of the United States is thought not to be applicable to modern agriculture; rather it is more likely to be used by the home gardener, by the organic gardener, or in areas of subsistence agriculture. These techniques must be reexamined. The various genetic methods of pest control either are largely unproven and highly theoretical schemes or require high technology and large sums of money to implement. Often the problems can be solved using biological control, a resistant variety, or a combination of other traditional methods—particularly integrated control centering around biological control, use of resistant varieties, cultural methods, and ecologically selective use of insecticides (only as really needed).

References

Anonymous. 1969. Insect management and control. Volume 3: *Principles of Plant and Animal Pest Control.* Nat. Acad. Sci. Publ. 1695. Washington, D.C. 508 pp.

Beck, S.D. 1965. Resistance of Plants to insects. *Annu. Rev. Entomol.* **10**:207–232.

David, D.E., R. Meyers, and J.B. Hoy. 1979. Biological control among vertebrates. In: C.B. Huffaker and P.S. Messenger (eds.) *Theory and Practice of Biological Control.* Academic Press: New York. pp. 501–519.

Knipling, E.F. and W.K. Klassen. 1976. Relative efficiencies of various genetic mechanisms for suppression of insect populations. U.S. Dept. Agric. Tech. Bull. 1533. 56 pp.

LaChance, L.E., C.H. Schmidt, and R.C. Bushland 1967. Radiation induced sterilization. In: W. W. Kilgore and R.L. Doutt (eds.) *Pest Control: Biological, Cultural, and Selected Chemical Methods.* Academic Press: New York. pp. 147–196.

Painter, R.H. 1951. *Insect Resistance in Crop Plants.* Macmillan: New York. 520 pp.

Painter, R.H. 1960. Breeding plants for resistance to insect pests. In: *Biological and Chemical Control of Plant and Animal Pests.* Am. Assoc. Adv. Sci: Washington D.C. pp. 245–266.

Robinson, R.A. 1976. *Plant Pathosystems.* Springer-Verlag: Berlin. 184 pp.

Roumasset, R.A. 1976. *Risk and Rice: Decision-making among Low Income Farmers.* North-Holland; Amsterdam. 251 pp.

Stern, V.M., R. van den Bosch, and T.F. Leigh. 1964. Strip cutting alfalfa for lygus bug control. *Calif. Agric.* **18**:406.

Thompson, H.V., and A.N. Worden. 1956. *The Rabbit.* Willmar Brothers. 240 pp.

Waterhouse, D.F. 1964. The biological control of dung. *Sci. Am.* **230**:100–109.

Whitten, M. 1970. Genetics of pests and their management. In: R.L. Rabb and F.E. Guthrie (eds.) *Concepts of Pest Management.* North Carolina State University Press: Raleigh. pp. 119–137.

12

Economics of Biological Controls

Importation of Biological Controls

Assessing the economic benefits and costs of imported biological control is difficult. Costs of research, quarantine, implementation, and overall organization are easy to measure, but many of the important benefits to agriculture and society are more difficult to quantify. The two obvious benefits of successful biological control projects may be seen in lower pest control costs to growers and increases in production. Even some partially successful projects (California red scale) requiring an occasional rerelease of natural enemies and the supplemental assistance of chemical and cultural controls may substantially reduce the total costs of control by 75% or more (DeBach, 1974). The total benefits to the ecosystem and the general public from lowered pesticide use are difficult to assess in monetary terms. What value can we place on a human life, a poisoned lake, contaminated groundwater, or future pesticide-induced cancers, mutations, or teratogenic effects? Taking a more practical view, biological control when successfully applied avoids many pesticide-induced problems such as pest resurgence, secondary pest outbreaks, phytotoxicity, pollinator mortality, pesticide resistance, and the above mentioned health problems. Clearly there is no easy way to estimate these benefits, but we all recognize that they are important. When an importation project is completely successful, the benefit accrues over future time, given continued production of one or more of the host crops.

The principal concept behind classical biological control is that, once an introduction has occurred, the natural enemy may fit so well into its new environment that it will completely control the target pest, and neither the natural enemy nor the pest will continue to occur in significant numbers. This is both the beauty and bane of this control technique. This method may be initially spectacular and is often observed as it occurs by the farmers and the public. Fortunately, the pest problem is reduced, but unfortunately, the farmer soon forgets about it. It no longer provides a lasting advertisment of biological control.

Classical California Examples

DeBach (1964) and Huffaker *et al*. (1976) have compiled rough estimates of the money saved by the California agricultural industry through use of various imported biological controls from 1928 to 1973 (Table 12.1). The combined savings attributed to seven major projects described below are in the neighborhood of $250 million, not accounting for inflation and/or discount. These estimates are believed to be conservative. One cause of understatement would be the lack of data on the extent to which biological controls have prevented enlargement of pest ranges.

1. Klamath Weed Control with Beetles. In 1944 over two million acres of rangeland in California were infested with the toxic Klamath weed, *Hypericum perforatum*. Use of the beetles *Chrysolina quadrigemina* and *C. hyperici* to control this weed greatly increased land values when the forage production

TABLE 12.1. *Estimates of Savings to the Agricultural Industry in California through Major Successful Biological Control Projects, 1928–1979*

Project	Degree of success	Yearly savings over previous losses plus pest control costs (× $1000)	Total savings to 1979
Klamath weed	Complete	$2000—weight gain in cattle (1953–1959)[a] $2500—weight gain in cattle (1959–1979)[b]	$ 66,500,000
Grapeleaf skeletonizer	Partial to complete	$75 (1945–1956)[a] $100 (1956–1979)[b]	$ 3,300,000
Spotted alfalfa aphid	Substantial	$5580 (1958–1959)[b] $3000 (1959–1979)[b]	$ 74,160,000
Citrophilus mealybug	Complete	$2000 (1930–1959)[a] $2500 (1959–1979)[b]	$ 92,500,000
Olive parlatoria scale	Complete	$465 (1962–1966)[b] $725 (1967–1979)[b]	$ 11,750,000
Black scale on citrus	Partial to complete	$1684 (1940–1959)[a] $2100 (1959–1979)[b]	$ 69,360,000
Walnut aphid Total	Substantial	$250 (1970–1979)[b]	$ 2,500,000 $987,070,000

[a]Drawn from Huffaker *et al*. (1976) cited from DeBach (1964). 1964 dollars.
[b]Computed by Huffaker *et al*. (1976) in 1973 dollars. Recalculated to 1979 dollars.

greatly increased and the limited herbicide "holding" program costs were eliminated. The savings in herbicide applications and revenue increases from cattle weight gains have greatly exceeded the $200,000–$300,000 expended during the importation of the two species of *Chrysolina* beetles (DeBach, 1974).

2. *Western Grapeleaf Skeletonizer Control*. Between 1945 and 1956 San Diego County spent $75,000 per year in an attempt to eradicate the grapeleaf skeletonizer, *Harrisinia brillians* B. and McD., with insecticides. Local authorities originally thought that total eradication of this grape pest could be achieved within 10 years, but this was not the case. Biological control, which was initiated in 1950, was originally assumed to have saved the County $750,000 (Table 12.1). The range of western grapeleaf skeletonizer has continued to increase, so much so that California Department of Food and Agriculture (CDFA) costs for an insecticide program for this pest during 1976 alone were $660,000. But, given the more recent realization that eradication is not a viable final solution under most circumstances, a greater emphasis has been placed on biological control, and the benefits are now viewed as accruing continuously over time.

4. *Citrophilus Mealybug Control*. The successful control of the citrophilus mealybug, *Pseudococcus fragilis*, by *Coccophagus gurneyi* and *Hungariella pretiosa* (Timb.) was completed within 2 years of the introductions. During years of high infestation, surveys showed annual losses of $500,000 to $1 million in Orange County citrus alone (DeBach, 1974). Entomologists believed the total infested citrus area to be at least twice as large as this known area in Orange County. The importation of the mealybug parasites, at a cost of less than $2600 in 1928 (DeBach, 1974), has resulted in gross savings to growers in excess of $90 million up to 1973 (Table 12.1).

5. *Olive Parlatoria Scale Control with Internal and External Parasites*. After its accidental establishment in Fresno olive orchards in 1934, olive parlatoria scale, *Parlatoria oleae*, spread through the Central Valley and into Southern California. Control costs and decreased olive production and quality reductions caused annual losses in excess of $1 million.

One imported external parasite, *Aphytis paramaculicornis*, reduced the scale population in many areas by 99%, but the residual population in some situations was still high enough to discolor and distort fruits. The additional foreign search for biological control agents produced the internal parasite, *Cocophagoides utilis*. The two parasites proved to be complementary and now provide almost total control (DeBach, 1974).

6. *Control of Black Scale on Citrus by Metaphycus Helvolus*. Before 1940, many Southern California citrus acres were fumigated with hydrogen cyanide gas or sprayed with oil every year in order to prevent losses caused by black scale, *Saissetia oleae*. Estimates by DeBach (1964) and Huffaker *et al*. (1976) of annual savings owing to the presence of the imported parasite

Metaphycus helvolus are extrapolated from the estimates of the Fillmore Citrus Protective District in Ventura County (Table 12.2 and 12.3). Though occasional inundative releases are necessary in some areas to achieve total control, savings were estimated to be approximately $40 per acre per year during 1960–1970 in the Fillmore District.

7. *Control of the Walnut Aphid by Parasitic Wasps.* In 1959, the highly specific parasitic wasp *Trioxys pallidus* was introduced from France for control of the alien walnut aphid, *Chromaphis juglandicola*. Though this particular wasp was a highly successful parasite in Southern California, weather conditions kept it from establishing in northern and central California. A later importation of an Iranian biotype of *Trioxys pallidus* during the 1970s brought about almost complete control in less than a year (DeBach, 1974; this text, Chapter 9), saving one-half of the potential losses owing to aphids on walnut trees from 1970 to the present. Annual savings due to the parasite are estimated at $350,000, based upon CDFA estimates of crop losses by the aphid during 1963–1969. Huffaker *et al* (1976) conservatively estimated annual savings of $750,000 during that time period. (Table 12.1).

Cost–Benefit Estimates

Each dollar expended for biological control by the University of California on importation, mass rearing, and release of natural enemies has resulted in a 25-fold return through prevention of yield or quality reduction and/or

TABLE 12.2. *Acres Treated in the Fillmore Crop Protective District with California Red Scale and Black Scale Parasites*[a]

	California red scale parasite		Black scale parasite	
Year	Acres treated (× 1000)	Releases per participant	Acres treated (× 1000)	Releases per participant
1969	10.7	1.24	15.1	1.75
1970	7.6	.90	13.5	1.60
1971	5.7	.68	12.6	1.52
1972	3.6	.45	12.2	1.54
1973	2.8	.34	19.3	2.35
1974	13.8	1.56	14.4	1.63
1975	14.2	1.58	14.2	1.58
1976	18.1	1.95	9.4	1.01
1977[b]				
1978[b]				

[a]Source: Fillmore Insectary Annual Reports (1970–1978).
[b]Not available on per acre basis.

TABLE 12.3. *Fillmore Citrus Protective District: Acres Treated with Chemical Controls for Various Citrus Pests as a Percent of Total Participant Acres*[a]

Year	Number of participant acres	Percent of acres treated				
		Black scale	Red scale	Aphids	Citrus thrips	Fall cleanup[b]
1962	8222	9.7	0	24.3	0.6	6.1
1963	8352	4.8	0	14.4	2.4	9.6
1964	8458	0	0	4.7	1.7	28.4
1965	8633	3.4	0.1	10.4	5.8	13.9
1966	8775	2.1	0.3	22.8[c]	10.8	45.6
1967	8783	0	0.3	17.1[c]	6.8	40.0
1968	8839	7.3	0.5	11.3[c]	10.2	33.9
1969	8628	3.6	0.3	9.3[c]	11.3	10.4
1970	8423	0.2	1.1	11.9[c]	9.5	23.7
1971	8335	0	0.3	6.0[c]	6.6	36.0
1972	7918	80.0[d]	0.3	3.8[e]	17.7	12.6
1973	8224	8.0	0.9	2.4[e]	9.7	25.6
1974	8811	1.3	4.0	12.0[e]	11.3	34.5
1975	9009	0	2.4	16.6[e]	4.3	27.6
1976	9296	0	0.2	9.1[e]	2.8	23.2
1977	9034	0	0	0	4.5	27.6
1978	9200	0	0	0	4.0	25.6

[a]Source: Fillmore Insectary Annual Reports (1970–1978).

[b]One to 1-2/5% light medium oil and micronutrient spray to prevent buildup of red spider mites, and to reduce red scale to controllable numbers so that parasites can "catch up."

[c]Includes micronutrients and/or 2,4-D.

[d]Occasional temperatures above 114°F during a few days in 1972 killed most black scale parasites, thus necessitating greater use of chemicals.

[e]Spring oil (aphid control) and minor elements.

reduction in control costs. DeBach (1964) and Huffaker and Messenger (1976) note, however, that many additional benefits have not been included in the assessment. The successes of imported biological control agents of nigra scale, *Saissetia nigra*, black scale, *Saissetia oleae*, and olive scale, *Parlatoria oleae*, on ornamentals; of alfalfa weevil, *Hypera postica,* and pea aphid, *Acyrthosiphon pisum*, on alfalfa; and of California red scale, *Aonidiella aurantii*, and yellow scale, *Aonidiella citrina*, on citrus are additional benefits. Still other examples of successful introductions include the importation of parasites for control of linden aphid, *Eucallipterus tiliae,* on linden, the control of prickly pear with coreid bugs and the cochineal scale, *Dactylopius ceylonicus,* on Santa Cruz Island, and of course the importation of the vedalia beetle, *Rodolia cardinalis*, for control of cottony-cushion scale, *Icerya purchasi*, on citrus. All of these successes are not included in the above estimates (Huffaker and Messenger, 1976, DeBach, 1964). The addition of these plus numerous others we have not listed that were complete or partial successes would greatly enlarge the benefits side of the equation.

Research and Development of Imported Biological Controls

As of 1976, some 1327 known successful or partially successful biological control introductions had occurred worldwide, with an estimated 81 of these reported as having occurred in at least one location in the U.S. (Laing and Hamai, 1976). Van den Bosch (1971) reports that the present annual rate of success of importation in the U.S. was 1.4 introductions per year over the last decade.

At this time the only *active* proponents of introduction and establishment of imported insects are *governments and universities*—the Universities of California and Florida, Texas A&M University, Virginia Polytechnic Institute, the State of Hawaii, and the federal government. Importation of predators and parasites by private industry or individuals is illegal because of the potential risks of introducing new pests or hyperparasites. Private industry should, however, be encouraged to fund the necessary research and exploration phases of biological control.

State and Federal Budgets for Imported Biological Controls

In 1976 the budget for foreign exploration at the University of California (UC), Riverside and Berkeley campuses, was only $8400 (Huffaker *et al.*, 1976). This is a paltry sum considering the benefit–cost ratio of past efforts. Most of the budget was primarily focused on the development of integrated control practices, which encourage the use of indigenous as well as imported natural enemy species (i.e., biological control makes integrated control possible). Clearly, more active support of classical biological control importations would greatly enhance integrated pest management programs, especially in light of the great economic success they have had. The California Department of Food and Agriculture (CDFA) has also supported several biological control projects, including the biological control of several weed species, western grape leaf skeletonizer, and the comstock mealybug, even though the Agency has not in the past been an active participant in biological control introductions. This latter activity in California was delegated to the University.

The Agricultural Research Service (ARS) of the United States Department of Agriculture (USDA) is responsible for importation of biological controls at the federal level. Between 1957 and 1975, approximately 140 scientific man-years and $2 million in supplemental research funds were expended for this purpose (National Academy of Sciences, 1975). Search for control of such key pests as the alfalfa weevil, *Hypera postica*; the cereal leaf beetle, *Oulema melanopus*; imported cabbageworm, *Pieris rapae*, alligator weed, *Alternanthera philoxeroides*; and halogeton weed, *Halogeton glomeratus* (May), accounted for approximately 50% of this effort. The Executive Committee of the National Academy of Sciences, studying the problems of pest control, has expressed concern over the low federal effort on classical biological control research. For example, during 1972 an average of 14 researchers per year were devoted to classical biological control, as compared to 679 devoted to all federal pest control efforts. This effort has been only slightly increased in recent years.

Costs of Introduction

Djerassi *et al.* (1974) write that the estimated costs of introduction of a biological control agent ranged from $127,150 at U.C. to $200,000 at USDA. Accounting for inclusion of the frequency of success rate, which he cites as 16 per 100 species at U.C., the cost per successful attempt ranged from $795,000 to $1.25 million. DeBach of U.C. Riverside, however, estimates that the chance of obtaining some significant extent of insect control from importation and permanent establishment of natural enemies to be about 54 in 100. DeBach (1974) writes that the research and development required to obtain a natural enemy may range from less than $100 to occassionally tens of thousands of dollars. The two estimates diverge because Djerassi *et al.* probably include the budget of the entire Division of Biological Control at U.C., whereas DeBach includes only the variable costs specifically associated with the travel and research and development phases of this type of effort. But whichever estimate is taken, the benefits to society far outweigh the investments.

Inundative Release of Imported or Native Natural Enemies and Environmental Enhancement

The use of inundative release programs of either imported or native natural enemies in conjunction with environmental manipulation (e.g., selective use of insecticides) to enhance the naturally occuring enemy complex has proven to be economically feasible for several crops in California. Dietrick (1972) claims that this combined approach reduced users' pest control costs by $125 per acre per year in citrus and by $50 to $70 per acre per year in cotton in the Central and Coachella Valleys of California, respectively, during 1972. These savings were largely attributed to the ability of these natural enemies to control pesticide-resistant pests. Similarly, the Fillmore Citrus Protective District (Table 12.2) claims substantial pest control savings to growers through a combination of inundative release and environmental manipulation.

Biological Control and the Public Sector

The private sector has largely restricted itself to insect production and inundative release of a relatively few imported and native natural enemies. Such a demand develops if the population of an *effective* foreign introduction or an indigenous biological control can adequately suppress the pest population within the season, but for some reason becomes too scarce between seasons as a result of severe winter conditions or low host densities. Privately sponsored cooperative projects or individual pest control consultants may mass-rear various species (e.g., *Trichogramma* spp., parasites of California red scale, black scale) and release them if need is indicated during monitoring. A natural enemy, however, is not covered by patent protection; consequently, colonization of these biological controls, except for use in glasshouses in Europe for the most part, has not offered adequate economic incentive to induce large corporations that are currently involved in chemical pesticide development to participate in these ventures.

The costs of development and maintenance of an insectary appear to be within reach of area cooperatives and small private enterprises (Fillmore Citrus Protective District, 1970–1976; Dietrick, 1972). Because expensive analytic equipment is not needed to determine residues or guarantee product quality, there are at present no costly registration requirements for mass-producing natural or imported beneficials. But this could change with the new EPA "Biorational Pesticide" regulations that are being formulated. However, at this time, entrance into this endeavor does not appear to be economically prohibitive. The problem exists in identifying and organizing the market for the natural enemies.

A Working Example of Mass Culture and Augmentation of Imported Parasites

The Fillmore Citrus Protective District (C. Farrar, personal communication, 1977) is a cooperative pest control association in Ventura County, California, that uses biological controls in an "integrated biological control program." Particular emphasis is placed on control of California red scale. *Aonidiella aurantii*, and black scale, *Saissetia oleae,* through occasional inundative releases of two parasites, *Aphytis melinus* and *Metaphycus helvolus,* that attack the two pests, respectively. Both parasites were originally introduced by the University of California. The district maintains its own insectaries and stresses minimal application of selective insecticides.

All participant orchards receive at least one release of parasites per year (Table 12.2). The Fillmore program does, however, pay attention to other pests. DeBach (1974) reports that eight other insects, which are considered major serious pests in surrounding districts or counties, are being effectively controlled in this district under the Fillmore scheme. This includes yellow scale, *Aonidiella citrina*; brown soft scale, *Coccus hesperidum* L., dictyospermum scale, *Chrysomphalus dictyospermi* (Morgan); cottony-cushion scale, *Icerya purchasi*; and four species of mealybugs. Once again, many of the controlling parasites are UC imports.

C. Farrar (personal communication, 1980) stated that successful California red and black scale control depended heavily upon effective ant control, as ants defend the scales from attack by the parasites. The District used chlordane as a soil ant control until its use was banned by Federal regulations. The substitute organophosphate material was not as effective and problems with the citrus mealybug, *Planococcus citri* (Risso), and cottony-cushion scale developed. In addition to poorer control, the costs of the new material were approximately three times that of chlordane. Chemical controls are utilized only when biological controls break down, and then they are limited to the areas of infestation. Broad-spectrum chemicals are avoided.

Further emphasis is placed upon proper timing. Nontoxic oil sprays, for example, are frequently used to reduce scale populations until the parasite–scale ratio is increased to the level where control by the parasite is assured. Citrus red mites also respond to the oil sprays that are applied only in the fall when mite pest–predator ratios indicate a need.

The percentage of areas treated annually with insecticides for black scale and red scale are 7.0% and 0.6% respectively (Table 12.3). Generally, less than 20% of all acres also need annual applications for aphids and thrips. Fall cleanup frequently includes a light medium-oil spray (with micronutrients included) to prevent the buildup of red spider mites and red scale. DeBach (1974) writes that most acres require less than one spray of insecticides per year in this district.

M. K. Lorbeer, past manager of the Fillmore project, estimated that biological control of the black scale alone saved participant growers an average of $23.74 per acre per year during 1940–1954 (DeBach 1964), while during 1960–1970 annual savings of $40 per acre were generated through use of all biological controls. Fruit quality was also maintained during these periods (DeBach, 1974). Each participant grower (10,000 acres total) was charged $13 per acre in 1980 (i.e., a net change of $3 in 6 years); which was used primarily to finance parasite production. In 1977, a new monthly inspection program was started for interested growers on approximately 800 acres at an annual extra charge of $4 per acre. These costs are substantially less than the $100+ per acre costs in other regions of California. Quite possibly, the monitoring program will not only help identify specific problem areas for parasite release, but also further reduce the need for pesticides.

Field programs such as this, and others such as the release of parasites of the greenhouse whitefly and of the predators of mites in glasshouse floriculture and vegetable production, are prime examples of the uses of biological control in integrated control projects. The use of commercial preparations of microbial pathogens is still another example. These methods show considerable promise, but further development is required and should be encouraged.

References

DeBach, P. 1964. The scope of biological control. In: P. DeBach (ed.) *Biological Control of Insect Pests and Weeds*. Chapman & Hall, Ltd.: London. p. 3–20.

DeBach, P. 1974. *Biological Control by Natural Enemies*. Cambridge University Press: New York. 323 pp.

Dietrick, E. J. 1972. Private enterprise pest management based on biological controls. *Proc. Tall Timbers Conf. Ecol. Anim. Contr. Habitat Manage.* **4**:7–20.

Djerassi, C., C. Shin-Coleman, and J. Deikman. 1974. Insect control of the future: Operational and policy aspects. *Science* **186**:596–607.

Fillmore Citrus Protective District. 1970–1976. Annual Reports. Ventura, California.

Huffaker, C. B., F. J. Simmonds, and J. E. Laing. 1976. In: C. B. Huffaker and P. S. Messenger (eds.) *Theory and Practice of Biological Control*. Academic Press: New York. p. 42–73.

Laing, J. E., and J. Hamai. 1976. Biological control of insect pests and weeds by imported parasites, predators, and pathogens. In: C. B. Huffaker and P. S. Messenger (eds.) *Theory and Practice of Biological Control*. Academic Press: New York. p. 685–743.

National Academy of Sciences. 1975. *Pest Control: An assessment of present and alternative technologies*. Vol. 1. National Academy of Sciences, Washington, D.C. 506 pp.

van den Bosch, R. 1971. Biological control of insects. *Annu. Rev. Ecol. Syst.* **2**:45–66.

13

The Future of Biological Control

Biological control, as measured by the permanent suppression of pest species, ranks as one of the most effective pest (insects, vertebrates, weeds, etc.) control tactics. No one really knows how much benefit has been derived from pest control effected by natural enemies. DeBach (1964) estimated that between 1923 and 1959 classical biological control introductions costing about $4.3 million benefited the California agroeconomy alone in the amount of about $115 million. Unquestionably, many more millions of dollars in direct benefits have accrued in the ensuing years. If we then add the benefits from programs carried out in California before 1923 and those from the many successful programs conducted elsewhere in the world, it is apparent that savings amounting to several hundreds of millions of dollars have been realized from classical biological control. And then, too, we must consider the contribution of biological control to human health and environmental quality in general.

But natural-enemy introduction programs are only partially responsible for the benefit realized from biological control; naturally occurring parasites, predators, and pathogens add immensely to the total. It is impossible to even guess at the total financial benefit realized from naturally occurring biological control, but, as was mentioned in Chapter 12, it is not inconceivable that this great natural force may well be crucial to our very survival.

What then of the future utilization and manipulation of natural enemies? Has all the cream been skimmed or can we expect even greater benefits from biological control? We take the optimistic view. Indeed, it is our opinion that, if expanded support and effort are forthcoming, even greater rewards can be expected, especially since this would permit unprecedented exploitation of naturally occurring biological control as well as greater efficiency in natural-enemy importations. The following discussion is a brief outline of our thoughts on the future trends and developments in biological control.

Classical Biological Control

Natural-enemy importation, which has been going on for more than 80 years, has hardly approached its full potential. Even in the United States, where for decades there has been a relatively high level of activity, such major pests as the codling moth, boll weevil, and green peach aphid have never been serious subjects of natural-enemy introduction efforts. However this is changing. In addition, there are scores of exotic pests of lesser importance against which there has been little effort in the area of natural-enemy introductions. For example, of the scores of pest aphid species in this country, almost all of which are exotic, only a handful have been targets of classical biological control.

There are many countries in the world where there has been virtually no effort in biological control. In some countries this neglect has been largely due to skepticism on the part of policy makers, while in others it has resulted from inadequate resources, both financial and technical. Some governments have met this latter problem through membership in the Commonwealth Institute of Biological Control, the International Organization of Biological Control, or in programs sponsored by the Food and Agricultural Organization (FAO) of the United Nations. But this participation requires financial commitment and such funds are not always available.

Even in those countries with the most active classical biological control programs, support has been far from lavish. For example, in the United States, USDA estimates of yearly federal expenditures for classical biological control against the whole spectrum of exotic pests probably amounts to a little more than $2.8 million ($1.2 million for insects and $1.6 million for weeds). This is to be contrasted with the massive amounts of federal money expended on eradication or area control programs against such pests as the fire ant ($10 million in 1971 alone), gypsy moth, or white fringed beetle, or the hundreds of millions of dollars that society expends yearly on chemical control of insects and the monitoring programs that the chemicals require.

But biological control's slim fiscal fare may well be in line for augmentation as increasing public concern over pesticides generates greater pressures for alternative controls. And if increased support is forthcoming, the numbers of biological control successes will certainly increase. As DeBach (1964, 1974) pointed out, the degree of success in classical biological control is directly proportional to the amount of effort given to natural-enemy introduction. In other words, those countries or agencies that have made the greatest effort have reaped the most benefit. Thus in the United States, Hawaii and California have had by far the greatest success in classical biological control largely because they have pursued this tactic much more vigorously than the other states. It would seem then that if DeBach's formula were applied on a global basis, and support for natural-enemy introduction was, say, doubled, there might well be a doubling of the number of successes. In fact, with our increased understanding of entomophagous insects and improved techniques in the selective procurement, production, and colonization of natural enemies, the success ratio might be substantially higher than what it has been historically. But whatever the relative degree of success, there is abundant opportunity to add significantly to the triumphs of classical biological control through increased effort. Society then must decide whether it wants to invest in such an expanded effort with its real potential for substantial economic and ecological benefit. In light of past success, it is difficult to envisage anything but increased public support for classical biological control in this era of ecological concern.

Naturally Occurring Biological Control

There can be little doubt that research on naturally occurring biological control will increase in the future. The proliferating integrated control programs with their heavy reliance on natural enemies is a clear indication of things to come (Huffaker, 1971, 1980). And as time passes and information and experience are gained, even better use will be made of naturally occurring biological control. As this occurs, a very important windfall will result, namely, increased knowledge of entomophagous arthropods. Thus, we can expect to gain continually expanding insight into the systematics, biology, ecology, physiology, ethology, host ranges, and nutrition of parasitic and predaceous arthropods, and an improved understanding of host–natural enemy interrelationships. This information, in turn, will have a postitive feedback effect in that it will increase even further our efficiency in the use of naturally occurring biological control.

Special Manipulation of Natural Enemies

Special manipulation of natural enemies can perhaps be termed the third dimension of biological control, and yet one of the most promising. Improved mass-propagation techniques, use of artificial diets, and better understanding of natural-enemy ecology are the basis for this optimism. Practices of particular promise are periodic colonization of natural enemies, their nutritional augmentation, and the development of artificial or cultural manipulations designed to retain and augment their populations in the crop environments.

Innovation in production techniques, utilization of artificial nutrients or other habitat improvements, precise timing of releases, and the use of more effective natural-enemy species or strains should all contribute importantly to improved efficacy in the periodic colonization tactic. Studies in England and

Europe (Hussey and Bravenboer, 1971; Greathead, 1976) have shown that there is an excellent potential for this technique in glasshouse pest control. Pest problems in a variety of high-value crops (e.g., strawberries, vegetables, ornamentals) and in parks and home gardens also seem particularly amenable to the periodic colonization technique. On the other hand, since this is an aspect of biological control that would involve the marketing of a product (natural enemies), it might well generate the kind of capitalization and production know-how that will permit profitable large-scale propagation of natural enemies. In fact, certain traditional pesticide-producing companies in Japan and Europe have already entered the natural-enemy production field on a limited scale.

Advances in the nutrition of entomophagous insects are being made along two lines. The first of these is the development of synthetic or semi-synthetic diets as substitutes for living hosts in the mass propagation of natural enemies. This use of synthetic nutrients may well be a key factor in the practical utilization of insectary-grown natural enemies as "biotic insecticides." Provision of foodstuffs for field populations of natural enemies is the second promising avenue in the nutrition area. Here, cheap but complete foodstuffs are sprayed on vegetation or otherwise presented to the natural enemies either to attract them to desired places, to sustain them during periods of natural food scarcity, or to increase their reproduction rates (Hagen and Bishop, 1979). In some cases all three goals are met by distribution in the field of a given nutrient. Our colleague K. S. Hagen has been a pioneer in this field, and already a dairy by-product, developed by him as a predator food supplement, is being used on a commercial basis in California.

Environmental manipulations to preserve and augment natural enemies have been little exploited. This perhaps derives from the prevailing ignorance of the true significance of naturally occurring biological control. But a drastic change is clearly in the offing. Studies on strip-harvesting of alfalfa in California leave little doubt that universal adoption of this practice in the state would strikingly reduce pest insect problems in this $200 million crop. The provision of nesting places, diapause quarters, nectar plants, and sheltering sites, and other tactics have all been effective enhancers of natural-enemy activity in a spectrum of crops (van den Bosch and Telford, 1964). This promising record is a clear challenge to expand our efforts in this area of natural-enemy augmentation.

Pest Management and Biological Control

In recent years, a broadened concept of pest control has emerged called *integrated pest management*. It is derived from the ecologically based integrated control approach to pest suppression (See Chapter 10), but it is a much-expanded concept that includes the multidisciplinary consideration of all types of pests, insects, plant pathogens, nematodes, and weeds whenever these are of concern in the same crop environment, as well as utilization of knowledge about the crop and the economics of crop production.

Because of the holistic framework, and because decisions made by or for the grower about control of insect pests, for example, are influenced by or will affect decisions made about plant disease or weed control, pest management programs will of necessity often rely on a computer-assisted approach for arriving at optimal programs and decisions. One result of computer-assisted systems analysis is the new practice, instituted to control the Egyptian alfalfa weevil, of shifting insecticide application when necessary to the mid-winter period, when most natural enemies are in a dormant stage, and in a location where they are less likely to be killed. Computer programs for calculating time-varying life tables of alfalfa growth and development, as well as the population dynamics of the weevil, are already available for use in the field. Many other such programs in other crops in California and elsewhere are completed or are nearing completion for field implementation. A new computer-based pest management network is currently under development in California.

Biological control will certainly be a major component of these pest management programs. This will be so not only because biological control is a commonly used component of integrated control, but also because the concept and philosophy of biological control conform with the concept of integrated pest management. Consequently, as the pest management idea becomes implemented through research and pilot developments, we can expect to see an expanding need for and interest in biological control.

Conclusion

Biological control, a natural phenomenon, is a great biotic force that helps regulate insect populations and those of a myriad of other organisms as well. Like so many of our natural resources, biological control can either be squandered, with detriment to us and our environment, or conserved, augmented, and manipulated with beneficial results.

We have it within our power to choose between these alternatives. The nature of that choice will have important bearing on our future success as a species on this planet.

References

DeBach, P. 1964. Successes, trends and future possibilities. In: P. DeBach (ed.) *Biological Control of Insect Pests and Weeds*. Chapman & Hall: London. pp. 673–713.

DeBach, P. 1974. *Biological Control by Natural Enemies*. Cambridge University Press: London. 323 pp.

Greathead, D. J. 1976. Pests of protected cultivation. In: D. J. Greathead (ed.) *A Review of Biological Control in Western and Southern Europe.*CIBC Tech. Comm. no. 7. Commonw. Agricult. Bureaux, Farnham Royal. pp. 52–64.

Hagen, K. S., and G. W. Bishop. 1979. Use of supplemental foods and behavioral chemicals to increase the effectiveness of natural enemies. In: D. W. Davis, S. C. Hoyt, J. A. McMurtry, and M. J. AliNiazee (eds.) *Biological Control and Insect Pest Management.* Univ. of Calif. Publ. Agric. Sci. no. 4096. pp. 49–60.

Huffaker, C. B. (ed.) 1971. *Biological Control.* Plenum Press: New York. 511 pp.

Huffaker, C. B. 1980. *New Technology of Pest Control.* Wiley–Interscience: New York. 500 pp.

Hussey, N.W., and L. Bravenboer. 1971. Control of pests in glasshouse culture by the introduction of natural enemies. In: C.B. Huffaker (ed.) *Biological Control.* Plenum: New York. pp. 195–216.

van den Bosch, R., and A. D. Telford. 1964. Environmental Modification and Biological Control In: P. DeBach (ed.) *Biological Control of Insect Pests and Weeds.* Chapman & Hall: London. pp. 459–488.

Appendix

List of Species Cited in the Text

Chapter	Scientific name (author*)	Common name†
4	*Acridotheres tristis* (L.)	Myna bird
3	*Acyrthosiphon kondoi* (Shinji)	Blue alfalfa aphid
3, 6, 9, 11	*Acyrthosiphon pisum* (Harris)	Pea aphid
6	*Adelges piceae* Ratz	Wooly balsam aphid
11	*Aedes aegypti* (L.)	Yellow fever mosquito
11	*Aedes scutellaris* (Walker)	Mosquito
5	*Aeschynomene virginica* (L.) B.S.P.	Northern jointvetch
9	*Agrilus hyperici* (Creutzer)	Klamath weed beetle
5	*Agrobacterium radiobacter* (*)	Competitor — *A. tumefaciens*
5	*Agrobacterium tumefaciens* (Sm. and Town, Conn.)	Para — fungal spores
9	*Agrypon flaveolatum* (Gravenhorst)	Para — winter moth
8	*Agathis diversus* (Mues.)	Para — *Grapholitha molesta*
9	*Aleurocanthus spiniferus* (Quaint.)	Spiny blackfly
3, 8, 9	*Aleurocanthus woglumi* Ashby	Citrus blackfly
4	*Alloxysta victrix* (Westw.)	2nd Para — *Aphidius* spp.
3, 9, 12	*Alternanthera philoxeroides* (Mart.) Gris.	Alligatorweed
10	*Amyelois transitella* (Walker)	Navel orange worm
9	*Anaphoidea nitens* Girault	Para — eucalyptus snout beetle
11	*Anasa tristis* (De Geer)	Squash bug
11	*Anastrepha ludens* (Loew)	Mexican fruit fly
10	*Anopheles albimanus* Weidemann	Mosquito — malaria vector
3, 9	*Anomala orientalis* Waterhouse	Anomala beetle
9	*Anomala sulcatula* Burm.	Oriental beetle
2, 8	*Anthonomus grandis* Boh.	Boll weevil
5	*Anticarsia gemmatalis* Hubner	Velvetbean caterpillar
9	*Antonina graminis* (Maskell)	Rhodesgrass mealybug
3, 6, 8, 9, 12	*Aonidiella aurantii* (Mask.)	Red scale
9, 12	*Aonidiella citrina* (Coq.)	Yellow scale
4	*Apanteles congregatus* (Say)	Para — hornworm larvae
3, 9	*Apanteles glomeratus* (L.)	Para — cabbage butterfly

225

Chapter	Scientific name (author*)	Common name†
9	*Apanteles rubecula* Marshall	Para — *Pieris rapae*
8	*Aphelinus asychis* Walk.	Para — spotted alfalfa aphid
7	*Aphelinus semiflavus* Howard	Para — spotted alfalfa aphid
3, 8	*Aphelinus mali* (Haldeman)	Para — wooly apple aphid
8	*Aphidius rubifolii* Makauer	Para — thimbleberry aphid
4	*Aphidius smithi* Sharma and Rao	Para — pea aphid
3	*Aphidius testaceipes* (Cresson)	Para — greenbug
6	*Aphis fabae* Scopoli	Bean aphid
9	*Aphis sacchari* Zhnt.	Sugarcane aphid
8	*Aphytis fisheri* DeBach	Para — red scale
8	*Aphytis lignanensis* (Masi)	Para — red scale
8, 9, 12	*Aphytis maculicornis* (Masi)	Para — olive scale
6, 12	*Aphytis melinus* DeBach	Para — red scale
9	*Aphytis paramaculicornis* (Masi)	Para — olive scale
5	*Apis melifera* L.	Honey bee
5	*Arachnula impatiens* Cienkowski	Amoeba — fungal spores, bacteria, nematodes, blue green algae
4	*Asaphes californicus* Girault	3rd Para — aphids
9	*Aspidotus destructor* Sign.	Coconut scale
6	*Aspidotus hederae* (Vallot)	Oleander scale
9	*Asterolecanium variolosum* (Ratz.)	Golden oak scale
5	*Autographa californica* (Speyer)	Alfalfa looper
5	*Azotobacter chroococcum* Beijerinck	Para — soil bacteria
1, 5	*Bacillus popilliae* Dutky	Milky disease — Japanese beetle
5	*Bacillus sphaericus* Meyer and Neide	Bacteria — mosquito larvae
5	*Bacillus subtilis* (Cohn) Praz.	Antibiotic bacteria
1, 5	*Bacillus thuringiensis* Berliner	Bacteria — lepidopterous larvae
5	*Bacillus thuringiensis israelensis*	Bacteria — mosquito larvae
6, 8	*Bathyplectes anurus* (Thompson)	Para — alfalfa weevil
8	*Bathyplectes curculionis* (Thompson)	Para — alfalfa weevil
5	*Bdellorvibrio bacteriovorus* Stolp & Starr	Para — gram negative bacteria
5	*Bdellovibrio starii* Ceid., Man., & Bap.	Para — gram negative bacteria
5	*Bdellovibrio stolpii* Ceid., Man., & Bap.	Para — gram negative bacteria
1, 5	*Beauveria bassiana* (Balsamo) Vuillemin	Fungi — several insect orders
5	*Beauveria tenella* (Delacroix) Siemaszko	Fungi — mosquito larvae
1	*Blissus leucopterus* (Say)	Chinch bug
5, 11	*Bombyx mori* L.	Silkworm
7	*Brevicoryne brassicae* L.	Cabbage aphid
9	*Brontispa longissima selebensis* Gestro	Coconut leaf miner
9	*Brontispa mariana* Spaeth	Mariana coconut beetle
10	*Bucculatrix thurberiella* Busck	Cotton leaf perforator
4, 11	*Bufo marinus* L.	Marine toad
5	*Cactoblastis cactorum* (Berg.)	Cactus moth
3	*Caliroa cerasi* (L.)	Pear slug
3, 4	*Callaphis juglandis* (Goeze)	Dusky-veined aphid
3	*Calosoma sycophanta* (L.)	Pred — carabid beetle

Chapter	Scientific name (author*)	Common name†
9	*Cavariella aegopodii* (Scop.)	Carrot aphid
3	*Cephus cinctus* Norton	Wheat-stem sawfly
9	*Cephus pygmaeus* (L.)	Wheat-stem sawfly
5	*Ceratocystis ulmi* (Buisman)	Dutch elm disease
2, 8, 9, 11	*Ceratitis capitata* (Wiedeman)	Mediterranean fruit fly
5	*Cercospora rodmanii* (*)	Fungi — water hyacynth
5	*Cercosporella agertinae* (*)	Fungi — Pamakani weed
9	*Ceroplastes rubens* Mask.	Red wax scale
5	*Chondrilla juncea* L.	Skeleton weed
2, 7	*Choristoneura fumiferana* (Clemens)	Spruce budworm
4, 6, 8, 9, 12	*Chromaphis juglandicola* (Kalt.)	Walnut aphid
9, 12	*Chrysolina hyperici* (Forster)	Klamath weed beetle
9, 12	*Chrysolina quadrigemina* Suffrian	Klamath weed beetle
4	*Chrysopa carnea* Stephens	Green lacewing
9	*Chrysomphalus aonidum* (L.)	Florida red scale
9, 12	*Chrysomphalus dictyospermi* (Morgan)	Dictyosperm scale
5	*Cleonus punctiventris* Germar	Sugarbeet curculio
9	*Clidemia hirta* L.	Curse
9	*Cnidocampa flavescens* (Wlk.)	Oriental moth
1	*Coccobacillus acridiorum* d'Herelle	Bact — grasshoppers
4, 9, 12	*Coccophagoides utilis* Doutt	Para — olive scale
12	*Coccophagus gurneyi* Comp.	Para — citrophilus mealybug
4	*Coccophagus scutellaris* Dalman	Para — scale
8	*Coccophagus* spp.	Para — scale
12	*Coccus hesperidium* L.	Soft brown scale
11	*Cochliomyia hominivorax* (Coqueral)	Screwworm fly
3, 9	*Coleophora laricella* Hubner	Larch case-bearer
5, 11	*Colias eurytheme* Boisduval	Alfalfa butterfly
5	*Colletotrichum globosum* (*)	Beneficial Fungi — soil microflora
5	*Colletotrichum gloeosporoides f. spp Aeschynomene* (Penz)	Fungi — Jointvetch
8	*Comperiella bifasciata* (Howard)	Para — Yellow scale
4	*Compsilura concinnata* Meigan	Tachinid Para — polyphagous
5	*Coniothyrium minitans* Campbell	Fungal para — Sclerotinia
3	*Conotrachelus nenuphar* (Herbst)	Plum curculio
11	*Contarinia sorghicola* (Coquillett)	Sorghum midge
3	*Convolvulus arvensis* (L.)	Bindweed (morninglory)
9	*Cordia macrostachya* (Jaquin)	—
5	*Cordyceps sobolifera* (Hill) Berk. and Br.	Fungi — cicada
3	*Cryptochetum iceryae* (Will.)	Para — cottony-cushion scale
6, 8	*Cryptolaemus montrouzieri* (Muls.)	Mealybug destroyer
11	*Culex pipiens pipiens* (L.)	Mosquito
10	*Culex tarsalis* Coquillett	Mosquito — encephalitis vector
5	*Culicinomyces* sp.	Fungi — mosquito larvae
9	*Cyzenis albicans* (Fallen)	Para — winter moth
5	*Dactylella oviparasitica* Stir. and Mank.	Para — root knot nematodes
12	*Dactylopius ceylonicus* Green	Cochineal scale vs prickly pear

Chapter	Scientific name (author*)	Common name†
11	*Dacus curcubitae* Coquillett	Melon fly
8, 9, 11	*Dacus dorsalis* Hendel	Oriental fruit fly
11	*Daktulosphaira (-Phylloxera) vitifoliae* (Fitch)	Grape phylloxera
9	*Dasyneura pyri* (Bouche)	Pear leaf midge
11	*Diabrotica* spp.	rootworms
7	*Diadromus plutellae* (Mues.)	Para — diamondback moth
8, 9	*Diatraea saccharalis* (F.)	Sugarcane borer
5	*Diceroprocta apache* Davis	Cicada
1	*Diprion hercyniae* (Hartig)	European sawfly
4	*Dytiscus* sp.	Predaceous diving beetle
3	*Eichornia crassipes* (Mart.) Solms	Water hyacinth
9	*Emex australis* Steinh.	Spiny emex
9	*Emex spinosa* (L.) Campd.	Spiny emex
5	*Entomophthora* spp.	Fungus-Insecta
3	*Ephialtes caudatus* (Ratzeburg)	Para — codling moth
8	*Epilachna varivestis* Muls.	Velvetbean caterpillar
8	*Eretmocerus serins* Silv.	Para — citrus black fly
8, 9	*Eriococcus coriaceus* (Mask.	Blue gum scale
3, 9	*Eriosoma lanigerum* (Hausm.)	Wooly apple aphid
2	*Erythroneura elegantula* Osborne	Grape leafhopper
3, 12	*Eucallipterus tilliae* (L.)	Linden aphid
3, 9	*Eulecanium coryli* L.	European fruit lecanum
9	*Eupatorium adenophorum* (Spreng.)	Pamakani
5	*Eupatorium riparia* Riley	Pamakani weed
11	*Euniticellus intermedius* L.	Dung beetle
11	*Fusarium* sp.	Fusarium wilt
4	*Gambusia affinis* Baird & Girard	Mosquito fish
11	*Garreta nitens* (*)	Dung beetle
4	*Geocoris pallens* Stal	Big-eyed bug
9	*Gilpinia hercyniae* Htg.	Spruce sawfly
11	*Glossina morsitans* Westw.	Tse-tse fly
11	*Glossina swynnertoni* Austen	Tse-tse fly
5	*Glugea pyraustae* (Paillot)	Microsporidian — European corn borer
9	*Gonipterus scutellatus* Gyll.	Eucalyptus snout beetle
2, 8	*Gossypium hirsutum* L.	Cotton
3, 8, 9	*Grapholitha molesta* (Busck)	Oriental fruit moth
12	*Halogeton glomeratus* (May)	Halogeton weed
9, 12	*Harrisinia brillians* B. and McD.	Western grape leaf skeletonizer
10	*Heliothis virescens* (Fabricius)	Tobacco budworm
5, 11	*Heliothis* sp.	—
1, 2, 5, 6, 10	*Heliothis zea*	Tobacco bollworm
6	*Hemiberlesia lataniae* Signoret	*Latania* scale
5	*Heterorhabditis* spp.	Nematode — lepidoptera and coleoptera
3, 4	*Hippodamia convergens* (Guerin.)	Convergent ladybeetle
5	*Hirsutella thompsoni* Fisher	Fungi — citrus rust mite
9	*Homona coffearia* Nietn.	Tea tortrix

Chapter	Scientific name (author*)	Common name†
7	*Horogenes insularis* (Cress.)	Para — diamondback moth pupa
12	*Hungariella pretiosa* (Timb.)	Para — citrophilus mealybug
3	*Hydrilla verticillata* (L.f.) Royle	Hydrilla
3	*Hypera brunneipennis* (Boh.)	Egyptian alfalfa weevil
3, 6, 8, 9, 11, 12	*Hypera postica* (Gyll.)	Alfalfa weevil
8	*Hyperecteina aldrichi* Mesn.	Para — Japanese beetle
9	*Hypericum calycerum* L.	St. Johnswort
3, 9, 12	*Hypericum perforatum* L.	Klamath weed
5	*Hyphantria cunea* (Drury)	Fall webworm
9	*Icerya aegyptiaca* Dougl.	Egyptian mealybug
9	*Icerya montserratensis* Riley and Howard	Fluted scale
3, 4, 9, 12	*Icerya purchasi* (Maskell)	Cottony-cushion scale
3	*Iridiomyrex humilis* Mayr	Argentine ant
9	*Ischnaspis longirostris* (Sign.)	Black thread scale
6	*Juglans regia* L.	Walnut tree
5	*Laetisaria* sp.	Para — soil fungus
9	*Lantana camara* var. *aculeata* (L.) Moldenke	Lantana
9	*Laspeyresia nigricana* (Steph.)	Pea moth
1, 3, 6, 11	*Laspeyresia pomonella* (L.)	Codling moth
9	*Lepiosaphes beckii* (Newm.)	Purple scale
3	*Leptomastidea abnormis* (Girault)	Para — citrus mealybug
9	*Levuana iridescens* B. B.	Coconut moth
3	*Lucilia sericata* Meigen	Sheep blowfly
4, 6, 10, 11	*Lygus hesperus* Knight	Lygus bug
3, 5	*Lymantria dispar* (L.) (see *Portheria*)	Gypsy moth
3	*Lyperosia irritans*	Horn fly
8	*Macrocentrus ancylivorus* Roh.	Para — *Grapholitha molesta*
6	*Macrosiphum euphorbiae* (Thomas)	Potato aphid
8	*Masonaphis maxima* (Mason)	Thimbleberry aphid
11	*Mayetiola destructor* (Say)	Hessian fly
6	*Medicago hispida* Haertner	Bur clover
6	*Medicago sativa* L.	Alfalfa
6	*Melilotus* spp.	Clovers
5	*Meloidogyne* spp.	Root-knot nemtodes
5	*Melolontha melolontha* L.	Common cockchafer
4	*Mesochorus* sp.	2nd Para — *Apanteles* sp.
8	*Mesoleius tenthredinis* Morley	Para — larch sawfly
5	*Metamassius hemipterous* (L.)	Banana weevil
2, 6, 8, 12	*Metaphycus helvolus* (Compere)	Para — citrus black scale
5	*Metarrhizium anisopliae* (Metch.) Sor.	Entomogenous fungus
3	*Metopolophium dirhodum* Walker	Cereal aphid
3	*Microlarinus lareynii* (*)	Herbivore — puncture vine
3	*Microlarinus lypriformis* (*)	Herbivore — puncture vine
7	*Microplitis plutellae* (Vier.)	Para — diamondback moth
2	*Monellia costalis* (Fitch)	Pecan aphid
5	*Morrenia odorata* (Hook et Arn.) Linley	Stranglervine

Chapter	Scientific name (author*)	Common name†
11	*Murgantia histrionica* (Hahn)	Harlequin cabbage bug
11	*Myzus persicae* (Sulzer)	Green peach aphid
5	*Neodiprion sertifer* (Geoffroy)	European pine sawfly
4	*Neoplectana glaseri* Steiner	Nematode — Japanese beetle
5	*Neoplectana* sp.	Nematode — Coleoptera and Lepidoptera
9	*Nezara viridula* (L.)	Green tomato bug
9	*Nipaecoccus nipae* (Mask.)	Avocado scale
5	*Nomuraea rileyi* (*)	Fungi — velvetbean caterpillar
5	*Nosema melolonthae* Kreig	Protozoan — common cockchafer
5	*Nosema locustae* Canning	Protozoan—grasshopper
3, 9	*Nygmia phaeorrhoea* (Donovan)	Brown-tail moth
3	*Oecophylla smaragdina* Fab.	Predatory ant
11	*Onthophagus gazella* (L.)	Dung beetle
9	*Operophthera brumata* (L.)	Winter moth
9	*Opius longicaudatus* (Ashmead)	Para — oriental fruitfly
9	*Opius oophilus* Fullaway	Para — oriental fruitfly
8	*Opius* spp.	Para — fruit flies
2	*Opius tryoni* Cameron	Para — Queensland fruitfly
9	*Opius vandenboschi* Fullaway	Para — oriental fruitfly
9	*Opuntia dilenii* (Ker. Gawl.) Haw.	Prickly pear
9	*Opuntia imbricata* (Haw.) D.D.	Walking-stick cholla
9	*Opuntia inermis* D.D.	Prickly pear
9	*Opuntia megacantha* Salm–Dyck	Mission prickly pear
9, 11, 12	*Opuntia* spp.	—
9	*Opuntia streptacantha* Lem	Prickly pear
9	*Opuntia stricta* Haw.	Prickly pear
9	*Opuntia triacantha* Sweet	Prickly pear
9	*Opuntia tuna* Mill.	Prickly pear
9	*Opuntia vulgaris* Mill.	Prickly pear
5	*Orgyia pseudotsugata* McDonnough	Douglas fir tussock moth
5	*Orius tristicolor* (White)	Pirate bug
9	*Orthezia insignis* Cougl.	Greenhouse Orthezia
3, 9	*Oryctes tarandus* (Oliv.)	Sugarcane rhinocerous beetle
5, 11	*Oryctolagus cuniculus* L.	European rabbit
3, 5	*Ostrinia nubilalis* (Hbn.)	European corn borer
3, 12	*Oulema melanopus* L.	Cereal leaf beetle
9	*Oxya chinensis* (Thumb.)	Chinese grasshopper
10	*Panonychus citri* (McGregor)	Citrus red mite
8	*Paradexoides epilachnae* Ald.	Para — Mexican beanbeetle
3, 8, 9, 12	*Parlatoria oleae* (Colvee)	Olive scale
9	*Patasson nitens* (Girault)	Para — Eucalyptus snout beetle
3, 6	*Pectinophora gossypiella* (Saunders)	Pink bollworm
5	*Penicillium oxalicum* (*)	Fungi — antibiotic microflora
5	*Peniophora gigantea* (Fr.) Mass.	Competitor — *Fomes annosus*
4	*Pentalitomastix plethoricus* Caltagirone	Para — navel orangeworm
3, 9	*Perkinsiella saccharicida* Kirk.	Sugarcane leafhopper

Chapter	Scientific name (author*)	Common name†
5	*Pharagmidium violaceum* (Schultz) Winter	Blackberry rust
9	*Phenacoccus aceris* Sign.	Apple mealybug
9	*Phenacoccus hirsutus* Green	Hibiscus mealybug
9	*Phenacoccus iceryoides* Green	none
5	*Phyllocoptruta oleivora* (Fishmead)	Citrus rust mite
3	*Phthorimaea operculella* Zeller	Potato tuberworm
11	*Phylloxera* (see *Daktulosphaira*)	—
5	*Phytophthora citrophthora* (Smith and Smith) Leon.	Fungi — stranglervine
11	*Phytophthora infestans* (Mont.) DeB.	Potato late blight
5	*Phytophthora palmivora* Butl.	Fungi — papaya root rot
9	*Piersis brassicae* (L.)	European cabbageworm
3, 9, 12	*Pieris rapae* (L.)	Cabbage butterfly
9	*Pinnaspis buxi* (Bouche)	none
8	*Planococcus citri* (Risso)	Citrus mealybug
9	*Planococcus kenyae* (LeP.)	Coffee mealybug
3	*Planococcus* spp.	Mealybugs
3, 7, 9	*Plutella maculipennis* (Curtis)	Diamondback moth
4	*Polistes apachus* Saussure	Predaceous wasp
1, 3, 5, 8, 11	*Popillia japonica* Newman	Japanese beetle
7, 8	*Praon exsoletum* Nees	Para — spotted alfalfa aphid
4	*Praon* sp.	Para — dusky-veined aphid
3, 8, 9	*Pristiphora erichsonii* (Htg.)	Larch sawfly
9	*Promecotheca papuana* Dziki	Coconut leaf mining beetle
9	*Promecotheca reichei* Baly	Coconut leaf mining beetle
8	*Prospatella perniciosi* Tower	Para — San Jose scale
9	*Pseudaulacaspis pentagona* Targ.	White peach scale
9	*Pseudococcus citriculus* Green	Green's mealybug
9	*Pseudococcus comstocki* (Kuw.)	Comstock mealybug
3, 12	*Pseudococcus fragilis* Brian	Citrophilus mealybug
9	*Pseudococcus gahani* Green	Citrophilus mealybug
9	*Pseudococcus* spp.	Mealybug
3	*Psila rosa* L.	Carrot rust fly
5	*Puccinia chondrillina* Bubak & Syd.	Fungi — skeleton weed
11	*Puccinia graminus* Persoon	Wheat-stem rust
3	*Pulvinaria delottoi* Gill	Iceplant scale (1 generation)
3	*Pulvinariella mesembryanthemi* (Vallot)	Iceplant scale (2 generations)
5	*Pythium oligandrum* Drechster	Para — other *Pythium* spp.
5	*Pythium ultimum* Trow	Seedling disease
9	*Quadraspidiotus perniciosus* Comst.	San Jose scale
3	*Rhabdoscelus obscurus* (Bvd.)	Sugarcane weevil
11	*Rattus* spp.	European rats
9	*Rhabdoscelus obscurus* (Bvd.)	Sugarcane weevil
8	*Rhagoletis completa* (Cresson)	Walnut husk fly
2	*Rhagoletis pomonella* (Walsh)	Apple maggot
8	*Rhagoletis* spp.	Fruit fly
5	*Rhizobium* spp.	Para — bacteria
3	*Rhizobius ventralis* Erickson	Pred — black scale

Chapter	Scientific name (author*)	Common name†
5	*Rhizoctonia solani* Kuehn	Damping off disease
3	*Rhopalosiphum maidis* Fitch	Cereal aphid
3	*Rhopalosiphum padi* L.	Cereal aphid
3	*Rhyacionia buoliana* Achiffermuller	Pine shoot moth
3, 4, 6, 12	*Rodolia cardinalis* (Mulsant)	Vedalia beetle
5	*Romanomermis culicivorax* Ross and Smith	Nematode — mosquito larvae
5	*Rubus constrictus* Mueller and Lefevre	Wild blackberry
8	*Rubus parviflorus* (Nutt.)	Thimbleberry
5	*Rubus ulmifolius* Schott	Wild blackberry
9, 12	*Saissetia nigra* (Nietn.)	Nigra scale
2, 3, 6, 8, 9, 12	*Saissetia oleae* (Olivier)	Citrus black scale
3	*Schizaphis graminum* (Rondani)	Green bug
5	*Sclerotinia sclerotia* (*)	Fungi — soil pathogen
5	*Sclerotinia sclerotiorum* (Lib.) DeB.	White mold fungus
5	*Sclerotium rolfsii* Sacc.	Root disease
5	*Scolytus multistriatus* Marsham	Elm bark beetle
3	*Scutellista cyanea* (Motschulsky)	Para — citrus black scale
9	*Senecio jacobeae* L.	Tansy ragwort
9	*Siphanta acuta* (Wlk.)	Torpedo bug planthopper
3	*Sirex* spp.	Sawflies
3	*Sitobion avenae* F.	Cereal aphid
3	*Sitodiplosis mosellana* (Gehin)	Wheat midge
11	*Solanum* sp.	Nightshades (potato, tomato)
1, 5, 10, 11	*Spodoptera exigua* (Hubner)	Beet armyworm
11	*Spodoptera praefica* (Grote)	Western yellow-striped armyworm
5	*Sporidesmium sclerotivorum* Veck., Ay, and Adams	Para — *Sclerotinia*
10	*Stethorus* sp.	Pred — citrus red mite
9	*Stilpnotia salicis* (L.)	Satin moth
3	*Sminthurus viridis* Lubb.	Lucerne flea
9	*Tarophagus proserpina* (Kirk.)	Taro leafhopper
11	*Teleogryllus commodus* (Walker)	Field cricket
3, 5, 6, 9, 8, 11	*Therioaphis trifolii* (Monell)	Spotted alfalfa aphid
8	*Tiphia* sp.	Para — Japanese beetle
3, 9	*Trialeurodes vaporariorum* (Westw.)	Greenhouse whitefly
9	*Tribulus cistoides* L.	Jamaica fever plant
3, 9	*Tribulus terrestris* L.	Puncturevine
5	*Trichoderma harzianum* Rifai	Para — soil fungi
5	*Trichoderma* sp.	Para — soil fungi
9	*Trichogramma evanescens* Westw.	Para — *Pieris rapae*
12	*Trichogramma* sp.	Para — Lepidoptera eggs
1, 5, 10	*Trichoplusia ni* (Hubner)	Cabbage looper
8	*Trichopoda pennipes* Fabricius	Green stink bug
6	*Trifolium* spp.	Clover
6	*Trigonella* spp.	Clover

Chapter	Scientific name (author*)	Common name†
9	*Trionymus saccharis* (Ckll.)	Pink sugarcane mealybug
4, 7, 8	*Trioxys complanatus* Quilis	Para — spotted alfalfa aphid
4, 6, 8, 9, 12	*Trioxys pallidus* (Haliday)	Para — walnut aphid
3	*Tyroglyphus phylloxerae* Riley	Pred mite — Grape phylloxora
3	*Tytthus mundulus* (Breddin)	Pred — sugarcane leafhopper
5	*Uredo eichorniae* Ciferri & Fragoso	Fungi — water hyacynth
11	*Vitis labrusca* Fox	American grape
11	*Vitis vinifera* L.	European grape
11	*Xenopus laveus* L.	African clawed frog
4	*Zelus renardii* Kolenati	Assassin bug predator

*, Author name unknown.

†Para, parasite or parasitoid; 2nd Para, secondary hyper parasite; 3rd Para, tertiary hyper parasite; Pred, predator.

Glossary

Adelphoparasitism Self-parasitism, that form of parasitism in which one sex develops parasitically in the body of the opposite sex.

Agroecosystem The ecosystem composed of cultivated land, the plants contained or grown thereon, and the animals associated with these plants.

Alien species An organism that has invaded and is growing in a new region.

Arrhenotoky That pattern or mode of reproduction in which progeny of both sexes are produced by mated females, the egg when unfertilized producing a viable, haploid male and when fertilized producing a viable diploid female.

Biological control The regulation of plant and animal numbers by biotic mortality agents (natural enemies); also, the use by man of natural enemies to control (reduce) the numbers of a pest animal or weed.

Biotic agents Living environmental factors that favor the well-being or bring about the premature death of animals and plants.

Broad-spectrum pesticide A material of such broad toxicity that it kills not only a range of pest species but also many nontarget species, such as natural enemies, honeybees, birds, and other forms of wildlife.

Carnivore An animal that feeds on other live animals.

Cleptoparasitism A case of multiple parasitism in which a parasitoid preferentially attacks a host already parasitized by another species rather than an unparasitized host.

Colonization The controlled release of a quantity of biological control agents in a favorable environment for the purpose of permanent or temporary establishment.

Community A characteristic assemblage of interacting populations of species in a particular habitat.

Competition The interference that occurs among individuals utilizing a common resource that is insufficient in quantity to satisfy the needs of all.

Consumer A heterotrophic organism or population, usually animal, fungus, or virus, that utilizes dead or living organic matter as food.

Cultural control A pest control method in which normal agronomic practices—tilling, planting, crop spacing, irrigating, harvesting, waste disposal, crop rotation—are altered so that the environment is less favorable or unfavorable for the pest.

Decomposer A heterotrophic organism that utilizes dead organic matter as food, decomposing it into more simple substances.

Delayed density-dependent mortality Mortality inflicted on the members of a population, the magnitude of which is determined by the density of the population at some time in the past.

Density-dependent mortality Mortality inflicted on the members of a population, the degree of which is related to or affected by the density of the population.

Density-independent mortality. Mortality inflicted on the members of a population, the degree of which is unrelated to or unaffected by the density of the population.

Deuterotoky That pattern or mode of parthenogenetic reproduction in which progeny of both sexes are produced by unmated females.

Direct density-dependent mortality Mortality inflicted on the members of a population, the magnitude of which increases as the current density of the population increases and which decreases promptly as the current density of the population declines.

Direct hyperparasite A parasitoid that searches out and deposits its egg in or on the body of its parasitic host, which may or may not be contained in the body of its own living host.

Economic threshold The density of a pest population below which it fails to cause enough injury to the crop to justify the cost of control efforts.

Ectoparasite A parasitoid that develops externally on the body of the host.

Encapsulation An immune response of certain insect hosts to nonadapted parasitoid eggs and young larvae wherein the immature parasitoids become covered with and eventually encysted by certain host blood cells, with death of the parasitoids soon following.

Endoparasite A parasitoid that develops within its host's body.

Entomophagous insect A predatory or parasitic insect that feeds on other insects.

Entomophagy The consumption of insects by other animals.

Epizootic The widespread or simultaneous occurrence of a disease in a large proportion of an animal population.

Establishment of natural enemy The permanent occurrence of an imported natural enemy in a new environment after the act of colonization has been carried out.

Exotic species An organism that evolved in one part of the world and that now occurs either accidentally or intentionally (i.e., through introduction by man) in a new region.

Factitious host An easily grown plant or animal species used as a host for the mass culture of a natural enemy in the insectary but which is not attacked by this enemy in nature.

Fecundity The number of eggs or offspring the females of a species can produce during their lifetime.

Food chain A trophic path or succession of populations through which energy flows in an ecosystem as a result of consumer–consumed relationships.

Food web A complex of branching, joining, or diverging food chains that connect together the various populations in an ecosystem.

Genetic control A pest control method that makes use of selected strains of the target species possessing chromosome aberrations or other genetic abnormalities such that, when released into the target population, mating with wild (normal) individuals results in production of sterile or less viable progeny.

Gregarious parasite A parasitoid whose food requirements are such that from several to many can develop simultaneously in or on the body of the host.

Habitat A physical portion of the environment within which a population is dispersed.

Haploid parthenogenesis The situation in which the unfertilized egg hatches and develops normally to produce a viable male adult whose cells contain only the haploid number of chromosomes.

Herbivore An animal that feeds on live green plants.

Host plant resistance A method of pest control in which crop varieties are used that are resistant, tolerant, or unattractive to the pest.

Host specificity The degree of restriction of the number of different plant or animal species that can serve as a food source for herbivorous or carnivorous species.

Hyperparasite An insect that is parasitic in or on another parasitic insect.

Indirect hyperparasite A secondary parasitoid that searches out and deposits its egg in the body of an unparasitized, nonparasitic host, the egg usually remaining undeveloped until the nonparasitic host is subsequently parasitized by a primary parasitoid, which then serves as host for the secondary.

Inoculative releases   The repeated colonization of relatively small numbers of a natural enemy for purposes of building up a population over several generations.

Integrated control   An ecologically based pest population management system that uses all suitable techniques to reduce or so manipulate the pest population that it is prevented from causing economically unacceptable injury to the crop.

Inundative release   A colonization of large numbers of a natural enemy for the immediate purpose of inflicting prompt mortality on the pest population.

Inverse density-dependent mortality   Mortality inflicted on the members of a population, the magnitude of which decreases as the current density of the population increases and which increases as the current density of the population decreases.

Mass culture   The propagation in an insectary of very large numbers of a biological control agent, often on a continuous basis over a period of months or years.

Microbial pesticide   A material composed of or containing pathogenic microbes or their toxic products for use in controlling a pest (e.g., insect, weed).

Monophagy   The restriction of an animal to the consumption of but one species of food organism.

Mortality factor   A factor or agency in the environment that causes the premature deah of an organism.

Multiple parasitism   The situation in which more than one parasitoid species occurs simultaneously in or on the body of the host.

Natality   The properties, reproductive or dispersive, that enable a population to increase in number.

Natural control   The collective action of environmental factors to maintain the numbers of a population within cetain upper and lower limits over a period of time.

Natural enemy   An animal or plant that causes the premature death of another animal or plant.

Niche   The place or position, in both a physical and a functional sense, of a species population in an ecosystem as determined by the full complex of environmental factors impinging on and limiting the population.

Nonreciprocal density-dependence   Mortality inflicted on a population by a biotic mortality factor whose own numbers are not changed as a consequence.

Oligophagy   The restriction of an animal to the consumption of a moderate range of food organisms.

Parasite   A small organism that lives and feeds in or on a larger host organism.

Parasitism   A biotic interaction involving a trophic relation between a small organism and its larger host.

Parasitoid   A parasitic insect that lives in or on and eventually kills a larger host insect (or other arthropod).

Parthenogenesis   The production of normal progeny by unmated females.

Pathogen   A microorganism that lives and feeds (parasitically) on or in a larger host organism and thereby causes injury to it.

Pathogenesis   The causing of a state of ill health, morbidity, and in some cases premature death, in a host organism by a pathogenic microorganism.

Periodic colonization   The frequent release of small or large numbers of a natural enemy for short-term control action rather than permanent establishment.

Pest   A species that because of its high numbers is able to inflict substantial harm on man, domesticated animals, or cultivated crops.

Pest resurgence   The rapid numerical rebound of a pest population after use of a broad-spectrum pesticide, brought about usually by the destruction of natural enemies that were otherwise holding the pest in check.

Phytophagy   The consumption of plants or plant tissues by animals.

Polyphagy   The consumption by an animal of a wide variety of food organisms.

Population   An aggregation of similar individuals in a continuous area that contains no potential breeding barriers.

Predation   A biotic interaction involving a trophic association between a large or strong animal and a small or weak animal.

Predator   An animal that feeds upon other animals (prey) that are either smaller or weaker than itself.

Primary parasite   A parasitoid that develops in or on nonparasitic hosts.

Producer   An autotrophic organism or population, usually of green plants, that procures energy from outside the ecosystem and through the process of photosynthesis converts this energy into living organic matter within the system.

Quarantine   The containment of an imported or immigrant individual in a special, escape-proof facility until its identity is determined and any associated, potentially hazardous organisms are eliminated.

Reciprocal density-dependence   Mortality inflicted on a population by a biotic mortality factor whose own numbers are changed as a consequence.

Regulation   As related to population dynamics, the control of population density.

Reproductive capacity   The capacity of the individuals in a population to increase in number by the production of progeny.

Scavenger   An animal that utilizes the dead bodies or tissues of other organisms as food.

Secondary parasite   A parasitoid that develops in or on a primary parasite.

Secondary pest outbreak   The rapid numerical increase to pest status of a scarce, noneconomic, phytophagous population after use of a broad-spectrum pesticide for control of another pest of the crop, brought about by the destruction of natural enemies that were otherwise holding the secondary pest in check.

Solitary parasite   A parasitoid whose food requirements are such that no more than one can develop successfully in the body of the host.

Stenophagy   The restriction of an animal to the consumption of but a narrow range of closely related species of food organisms.

Sterile-insect control   A pest control method that makes use of artificially sterilized populations of the pest to mate with and thereby interfere with the normal reproductive efforts of the target species.

Superparasitism   The situation in which more individuals of a parasitoid species occur in a host than can survive.

Tertiary parasite   A parasitoid that develops in or on a secondary parasite.

Thelyotoky   That pattern or mode of parthenogenetic reproduction in which unmated females produce only female progeny, males being unknown.

Trophic level   A particular step occupied by a species population in the process of energy transfer within an ecocystem.

Weed   An aggressive, invasive, easily dispersed plant, one which commonly grows in cultivated ground to the detriment of a crop.

Index